THE BEST PRE-GED Study Series

INTERPRETING LITERATURE & THE ARTS

Elizabeth L. Chesla, M.A.
Polytechnic University, Brooklyn, NY

Research & Education Association
61 Ethel Road West, Piscataway, NJ 08854

The Best PRE-GED Study Series
INTERPRETING LITERATURE & THE ARTS

Copyright © 1998, by Research & Education Association. All rights reserved. No part of this book may be reproduced in any form without permission of the publisher.

Printed in the United States of America

Library of Congress Catalog Card Number 97-67680

International Standard Book Number 0-87891-797-7

Research & Education Association
61 Ethel Road West
Piscataway, New Jersey 08854

TABLE OF CONTENTS

Introduction v
    About this Book vii
    About the GED vii
    How to Use this Book viii
    About Research and Education Association ix
    Acknowledgments ix

Pre-Test 1
    Pre-Test 3
    Answer Key 12
    Pre-Test Self-Evaluation 13
    Answers and Explanations 15

Reading Literature 21
    What *is* Literature (A Working Definition) 23
    Reading Literature: A Search for Meaning 23
    Pre-Reading 24
    Active Reading 24
    From Active Reading to Comprehension: Making Observations and Drawing Inferences ... 27
    Review 37

Reading Prose 39
    Defining Prose: Fact and Fiction 41
    Fiction 41
    Elements of Fiction 42
    Major Movements in Fiction 46
    Non-Fiction 58
    Elements of Non-Fiction 58
    The Essay 60
    Biography and Autobiography 64
    Journalism 67
    Evolution of Non-Fiction 68
    Review 72

Reading Poetry ... **75**
What is Poetry? ... 77
Elements of Sound ... 77
Elements of Sense ... 79
Poetic Forms: The Importance of Structure ... 83
Major Movements in Poetry ... 89
Review ... 94

Reading Drama ... **95**
Defining Drama ... 97
Elements of Drama ... 101
A Brief History of Drama ... 103
Review ... 112

Reading Commentary ... **115**
What is Commentary? ... 117
Book Reviews ... 118
Theater Reviews ... 120
Music Reviews ... 121
Movie and TV Reviews ... 123
Review ... 128

Post-Test ... **129**
Post-Test ... 131
Answer Key ... 141
Post-Test Self-Evaluation ... 142
Answers and Explanations ... 144

Appendix A: Glossary of Terms ... **151**

Appendix B: Prefixes, Suffixes, and Word Roots ... **159**

Interpreting Literature
and the Arts

Introduction

ABOUT THIS BOOK

This book is designed to help you strengthen the reading comprehension skills you will need to take the "Interpreting Literature and the Arts" test of the **General Educational Development (GED) Examination**.

A **Pre-Test** section in the beginning of this book will help you assess the areas where you need to work the hardest. After you have completed the review areas and answered all of the drill questions, you will be given a **Post-Test** that will show your improvement in certain areas and show you which areas you still need to study. In the Post-Test section you will answer questions very similar to those you will face on the actual GED. The Pre-Test and Post-Test provide detailed explanations to all of the questions, illustrating not only why the correct answer choice was right, but also why the incorrect answer choices were wrong.

The reviews cover all areas tested on the Interpreting Literature and the Arts test of the GED examination. The four sections of Prose, Poetry, Drama, and Commentary are covered extensively and provide a hands-on approach to understanding the material. These reviews will help you process information, find main ideas, and remember important details. By mastering the skills presented in this book, you will be able to approach any work of literature with confidence.

ABOUT THE GED

The GED is an examination for adults who did not complete high school and would like to earn a high school equivalency diploma. The exam is given by each state, which then issues a GED diploma. The GED is taken by adults who want or need a diploma for work, college, or personal satisfaction. Nearly 700,000 people take the GED each year.

The GED is broken into five tests: **Writing Skills; Social Studies; Science; Interpreting Literature and the Arts; and Mathematics**. You are given about seven and a half hours to complete the examination. There are a total of 286 multiple-choice questions and an essay question on the GED examination.

The Interpreting Literature and the Arts test on the GED contains 45 multiple-choice questions to be answered in 65 minutes. The questions in this test are based on excerpts from poetry, essays, biographies, plays, critical reviews (of television, film, literature, dance, art, music, sculpture, and theater), and commentary. This test consists of 50 percent popular literature questions, 25 percent classical literature questions, and 25 percent commentary on literature and the arts questions. In this test, 60 percent of the questions will deal with comprehension, which will ask you to restate ideas or information presented in the passage or summarize what you have read. Application questions make up 15 percent of the questions and will ask you to draw your own conclusions, explain the effects or importance of a situation presented in a passage, or to identify the implications involved in a situation presented in a passage. The test also consists of 25 percent analysis questions which will ask you to define stylistic and structural techniques in terms of concept.

The GED examination is administered by the GED Testing Service of the American Council of Education (ACE) and is developed by writers who have secondary and adult education experience. Because the GED test-takers come from such diverse backgrounds, the ACE makes sure the test writers are also diverse. Once the questions have been written, they are standardized according to a certain level of difficulty and content.

The GED comes in several versions to fit the special needs of its examinees. For example, there are Spanish, French, and Braille versions of the exam, as well as large-print and audio versions.

If you would like to obtain more information about the GED, such as when and where it is administered, contact your local high school or adult education center. You can also call or write to the GED Testing Service at:

1-800-62MYGED
(1-800-626-9433)
General Educational Development
GED Testing Service
American Council on Education
One Dupont Circle, NW
Washington, D.C. 20036

HOW TO USE THIS BOOK

Before you begin reading the chapters on grammar, take the Pre-Test at the front of this book. This Pre-Test will assess your current reading comprehension skills and indicate both your strengths and weaknesses.

This book is broken down into four sections, each dealing with a specific genre of literature: prose (general literature), poetry, drama, and commentary. Each section includes several exercises to help you develop your reading skills as well as skill-building practice exercises at the end of each section to reinforce what you have learned.

It is best to move through this book from front to back because the Introduction to Reading Literature section lays the foundation for the rest of the book. In addition, the Fiction sub-section of the Reading Prose section discusses elements of literature that will recur throughout the book.

Because reading comprehension is a skill that is developed and not simply a list of facts to learn, it is important to begin developing this skill as soon as possible. Cramming for this test on the GED simply will not help. It is also wise not to try to do *too* much in one study session.

There is a large amount of information presented in this book, and it will take time to digest.

When you are finished with the review sections, take the Post-Test. Compare your Post-Test score with your Pre-Test score and see how much you have improved. You may even want to take the Pre-Test again to re-evaluate your skills.

Before you begin the Pre-Test, you should take inventory of your study skills. Under what circumstances do you study best? Are you most awake in the morning, late afternoon, or evening? Do you study best under bright light or soft? With music, or in total silence? Answer these questions and then try to optimize your study time by creating the ideal learning conditions (the atmosphere in which you learn most efficiently). Also, you should set a specific schedule for yourself that takes into consideration your other commitments. Can you study one hour each morning? A half hour every night? Two or three hours on weekends? Pick a routine and stick to it so you will be confident when it's time to take the exam.

Although you are encouraged to write in the margins of this book, you should know that you will not be allowed to mark up the texts on the actual GED examination. When you take the Pre-Test and Post-Test, use a piece of scrap paper instead of writing in the margins of the book. This will help you be more comfortable with the actual test format.

Please note that the GED won't penalize you for guessing wrong, so if you are stuck on a question, don't leave it blank. Instead, eliminate any answers you know are not correct and choose one of the remaining options. You have a better chance of getting a question right by making an educated guess, and then you can move on to the next question. Remember, the exam is timed, so you do not want to spend too much time on any one question.

ABOUT RESEARCH AND EDUCATION ASSOCIATION

Research and Education Association (REA) is an organization of educators, scientists, and engineers who specialize in various academic fields. REA was founded in 1959 for the purpose of disseminating the most recently developed scientific information to groups in industry, government, high schools, and universities. Since then, REA has become a successful and highly respected publisher of study aids, test preps, handbooks, and reference works.

REA's publications and educational materials are highly regarded for their significant contribution to the quest for excellence that characterizes today's educational goals. We continually receive an unprecedented amount of praise from professionals, instructors, librarians, parents, and students for our books. Our authors are as diverse as the subjects and fields represented in the books we publish. They are well-known in their respective fields and serve on the faculties of prestigious universities throughout the United States.

ACKNOWLEDGMENTS

Special recognition is extended to the following persons:

Dr. M. Fogiel, President, for his overall guidance which has brought this publication to completion.

Stacey A. Sporer-Daly, Managing Editor, for directing the editorial staff throughout each phase of the project.

Judy Winchock for editing the manuscript for technical accuracy.

Chris Dickinson, Kevin Gilligan, and Cheryl Pedersen for their editorial contributions.

Marty Perzan for typesetting the manuscript.

Interpreting Literature and the Arts

Pre-Test

INTERPRETING LITERATURE AND THE ARTS

PRE-TEST

DIRECTIONS: Read each of the passages below and then answer the questions pertaining to them. Choose the <u>best answer choice</u> for each question.

Questions 1–5 are based on the following passage.

PASSAGE A

　　I went to the woods because I wished to live deliberately, to front only the essential facts of life, and see if I could not learn what it had to teach, and not, when I came
(5) to die, discover that I had not lived. I did not wish to live what was not life, living is so dear; nor did I wish to practise resignation, unless it was quite necessary. I wanted to live deep and suck out all the
(10) marrow of life, to live so sturdily and Spartan like as to put to rout all that was not life, to cut a broad swath and shave close, to drive life into a corner, and reduce it to its lowest terms, and, if it proved to be
(15) mean, why then to get the whole and genuine meanness of it, and publish its meanness to the world; or if it were sublime, to know it by experience, and be able to give a true account of it in my next excursion.
(20) For most men, it appears to me, are in a strange uncertainty about it, whether it is of the devil or of God, and have *somewhat hastily* concluded that it is the chief end of man here to "glorify God and enjoy him
(25) forever."
　　Still we live meanly, like ants; though the fable tells us that we were long ago changed into men; like pygmies we fight with cranes; it is error upon error, and
(30) clout upon clout, and our best virtue has for its occasion a superfluous and evitable wretchedness. Our life is frittered away by detail. An honest man has hardly need to count more than his ten fingers, or in ex-
(35) treme cases he may add his ten toes, and lump the rest. Simplicity, simplicity, simplicity! I say, let your affairs be two or three, and not a hundred or a thousand; instead of a million count a half a dozen, and
(40) keep your accounts on your thumb-nail.

From "Walden" by Henry David Thoreau (1854)

1. The author went to the woods to
   (1) discover the meaning of life.
   (2) be alone.
   (3) write.

2. According to the author, what is wrong with our lives?
   (1) Not enough honest people
   (2) Too many details and distractions
   (3) Too much fighting

3. The italicized phrase "somewhat hastily" indicates that the author
   (1) believes this is the correct conclusion.
   (2) believes that there is no answer.
   (3) believes that this is not the chief end of life.

4. This passage is
   (1) poetry.
   (2) commentary.
   (3) non-fiction.

5. The author's purpose in this passage is to
   (1) explain.
   (2) entertain.
   (3) convince.

Questions 6–10 are based on the following passage.

PASSAGE B

Finally all the young people had gone to the altar and were saved, but one boy and me. He was a rounder's son named Westley. Westley and I were surrounded by
(5) sisters and deacons praying. It was very hot in the church, and getting late now. Finally Westley said to me in a whisper: "God damn! I'm tired o'sitting here. Let's get up and be saved." So he got up and was
(10) saved.

Then I was left all alone on the mourners' bench. My aunt came and knelt at my knees and cried, while prayers and songs swirled all around me in the little
(15) church. The whole congregation prayed for me alone, in a mighty wail of moans and voices. And I kept waiting serenely for Jesus, waiting, waiting—but he didn't come. I wanted to see him, but nothing
(20) happened to me. Nothing! I wanted something to happen to me, but nothing happened.

I heard the songs and the minister saying "Why don't you come? My dear
(25) child, why don't you come to Jesus? Jesus is waiting for you. He wants you. Why don't you come? Sister Reed, what's this child's name?"

"Langston," my aunt sobbed.

(30) "Langston, why don't you come? Why don't you come and be saved? Oh, Lamb of God! Why don't you come?"

Now it was really getting late. I began to be ashamed of myself, holding ev-
(35) erything up so long. I began to wonder what God thought about Westley, who certainly hadn't seen Jesus either, but who was now sitting proudly on the platform, swinging his knickerbockered legs and
(40) grinning down at me, surrounded by deacons and old women on their knees praying. God had not struck Westley dead for taking his name in vain or for lying in the temple. So I decided that maybe to save
(45) further trouble, I'd better lie too, and say that Jesus had come, and get up and be saved.

So I got up.

From "Salvation," by Langston Hughes (1940)

6. Why did Westley get up to be saved?

   (1) Because his mother told him to

   (2) Because he got tired of waiting

   (3) Because Hughes got up

7. What was the author waiting for?

   (1) For Jesus to show himself

   (2) For his aunt to help him

   (3) For the minister to talk to him

8. The author got up because

   (1) he had to go to the bathroom.

   (2) he saw Jesus.

   (3) he was making everyone wait.

9. "So I got up" is in a paragraph by itself because

   (1) the author only imagined getting up.

   (2) it is what everyone was waiting for.

   (3) he finally saw Jesus.

10. We can conclude that the author

    (1) was misled about what would happen in church.

    (2) was very religious.

    (3) was very good friends with Westley.

Questions 11–15 are based on the following excerpt.

PASSAGE C

BIFF: I tell ya, Hap, I don't know what the future is. I don't know—what I'm supposed to want.

HAPPY: What do you mean?

(5) . . .

BIFF: [*with rising agitation*]: Hap, I've had twenty or thirty different kinds of jobs since I left home before the war, and it always turns out the same. I just (10) realized it lately. In Nebraska, when I herded cattle, and the Dakotas, and Arizona, and now in Texas. It's why I came home now, I guess, because I realized it. This farm I work on, it's spring there (15) now, see? And they've got about fifteen new colts. There's nothing more inspiring or—beautiful than the sight of a mare and a new colt. And it's cool there now, see? Texas is cool now, and it's (20) spring. And whenever spring comes to where I am, I suddenly get the feeling, my God, I'm not gettin' anywhere! What the hell am I doing, playing around with horses, twenty-eight dol-(25) lars a week! I'm thirty-four years old, I oughta be makin' my future. That's when I come running home. And now, I get here, and I don't know what to do with myself. [*After a pause.*] I've al-(30) ways made a point of not wasting my life, and every time I come back here I know that all I've done is to waste my life.

HAPPY: You're a poet, you know that, (35) Biff? You're a—you're an idealist!

BIFF: No, I'm mixed up very bad. Maybe I oughta get married. Maybe I oughta get stuck into something. Maybe that's my trouble. I'm like a boy. (40) I'm not married, I'm not in business, I just—I'm like a boy. Are you content, Hap? You're a success, aren't you? Are you content?

HAPPY: Hell, no!

(45) BIFF: Why? You're making money, aren't you?

HAPPY: [*moving about with energy, expressiveness*] All I can do now is wait for the merchandise manager to die. (50) And suppose I get to be merchandise manager? He's a good friend of mine,

and he just built a terrific estate on Long Island. And he lived there about two months and sold it, and now he's
(55) building another one. He can't enjoy it once it's finished. And I know that's just what I would do. I don't know what the hell I'm workin' for. Sometimes I sit in my apartment—all alone. And I think
(60) of the rent I'm paying. And it's crazy. But then, it's what I always wanted. My own apartment, a car, and plenty of women. And still, goddammit, I'm lonely.

From *Death of a Salesman* by Arthur Miller (1949)

11. Biff is home because

    (1) he is making money.

    (2) he only gets $28 a week.

    (3) he isn't getting anywhere in life.

12. Biff thinks that Happy is happy because

    (1) he has a job.

    (2) he has a lot of girlfriends.

    (3) he has an apartment and a car.

13. Happy can best be described by which sentence?

    (1) You can't run away from yourself.

    (2) Money can't buy you happiness.

    (3) What goes around comes around.

14. Why isn't the merchandise manager happy?

    (1) He doesn't have enough money.

    (2) Happy is after his job.

    (3) The more he has, the more he wants.

15. We can conclude from this passage that Happy's name is

    (1) not his real name.

    (2) symbolic.

    (3) ironic.

Questions 16–22 refer to the following poem.

PASSAGE D

DIVORCING

One garland
of flowers, leaves, thorns
was twined round our two necks.
Drawn tight, it could choke us,
(5) yet we loved its scratchy grace,
our fragrant yoke.
We were Siamese twins.
Our blood's not sure
if it can circulate,
(10) now we are cut apart.
Something in each of us is waiting
to see if we can survive,
severed.

By Denise Levertov (1970)

16. Which adjective best describes how the speaker feels about the divorce?

    (1) Happy

    (2) Indifferent

    (3) Empty

17. Line 4 suggests that

    (1) marriage can strangle love.

    (2) someone tried to choke the speaker.

    (3) marriage is like a fancy necklace.

18. "Siamese twins" in line 7 is

    (1) personification.

(2) a metaphor.

(3) a simile.

19. "Yoke" in line 6 suggests that

   (1) marriage requires work.

   (2) marriage is like an egg.

   (3) their marriage is a burden.

20. This poem is

   (1) free verse.

   (2) blank verse.

   (3) a sonnet.

21. Which sentence best sums up the relationship between the speaker and spouse?

   (1) They had become too selfish.

   (2) They had become like one person.

   (3) They were very patient.

22. The divorce

   (1) will be good for them.

   (2) will never happen.

   (3) may kill them.

Questions 23–27 refer to the following passage.

PASSAGE E

She was one of those pretty and charming girls who are sometimes, as if by a mistake of destiny, born in a family of clerks. She had no dowry, no expectations, (5) no means of being known, understood, loved, wedded by any rich and distinguished man; and she let herself be married to a little clerk at the Ministry of Public Instructions.

(10) She dressed plainly because she could not dress well, but she was as unhappy as though she had really fallen from her proper station, since with women there is neither caste nor rank: and beauty, grace, (15) and charm act instead of family and birth. Natural fineness, instinct for what is elegant, suppleness of wit, are the sole hierarchy, and make from women of the people the equals of the very greatest la- (20) dies.

She suffered ceaselessly, feeling herself born for all the delicacies and all the luxuries. She suffered from the poverty of her dwelling, from the wretched look of (25) the walls, from the worn-out chairs, from the ugliness of the curtains. All those things, of which another woman of her rank would never even have been conscious, tortured her and made her angry. (30) The sight of the little Breton peasant, who did her humble housework aroused in her regrets which were despairing, and distracted dreams. She thought of the silent antechambers hung with Oriental tapestry, (35) lit by tall bronze candelabra, and of the two great footmen in knee breeches who sleep in the big armchairs, made drowsy by the heavy warmth of the hot-air stove. She thought of the long *salons* fitted up (40) with ancient silk, of the delicate furniture carrying priceless curiosities, and of the coquettish perfumed boudoirs made for talks at five o'clock with intimate friends, with men famous and sought after, whom (45) all women envy and whose attention they all desire.

From "The Necklace," by Guy De Maupassant

23. The woman described in this passage is

   (1) a member of the nobility.

   (2) a member of the working class.

   (3) a servant.

24. The phrase "she let herself be married to a little clerk" suggests that
    (1) she let her husband do whatever he wanted.
    (2) she thought she was above marrying him.
    (3) she'd always wanted to marry a clerk.

25. The word "ceaselessly" in paragraph 3 means
    (1) occasionally.
    (2) deservedly.
    (3) constantly.

26. She differed from other women of her rank because
    (1) she felt she deserved a higher rank.
    (2) she married a clerk.
    (3) she had a taste for elegance.

27. She says a woman's social rank
    (1) is not as important as a man's.
    (2) can change, but a man's can't.
    (3) should be determined by beauty and charm, not birth.

<u>Questions 28–30</u> refer to the following passage.

PASSAGE F

It was at this period of my emerging from the vast solitude in which I had been making my own acquaintance, that I
(5) stumbled upon Maupassant. I read his stories and marvelled at them. Here was life, not fiction; for where were the plots, the old-fashioned mechanism and stage trapping in that vague, unthinking way I had fancied were essential to the art of story
(10) making? Here was a man who had escaped from tradition and authority, who had entered into himself and looked out upon life through his own being and with his own eyes; and who, in a direct and simple way,
(15) told us what he saw. When a man does this, he gives us the best that he can; something valuable for it is genuine and spontaneous. He gives us his impressions. Someone told me the other day that
(20) Maupassant had gone out of fashion. I was not grieved to hear it. He has never seemed to me to belong to the multitude, but rather to the individual. He is not one whom we gather in crowds to listen to—whom we
(25) follow in procession—with beating of brass instruments. He does not move us to throw ourselves in the throng—having the integral of an unthinking whole to shout his praise. I even like to think that he ap-
(30) peals to me alone.

From "How I Stumbled upon Maupassant," by Kate Chopin (1896)

28. The writer found Maupassant's stories so moving because
    (1) they were written just for her.
    (2) they were written from a new perspective.

(3) it was fashionable to read them.

29. "Here was life, not fiction" suggests that

    (1) Maupassant was a biographer.

    (2) Maupassant was a reporter.

    (3) Maupassant was a realist.

30. The writer says Maupassant "gives us the best that he can" because

    (1) this is the best he can write.

    (2) he practiced his technique for a long time.

    (3) he shares something of himself in his fiction.

Questions 31–36 refer to the following poem.

PASSAGE G

A POISON TREE

I was angry with my friend:
I told my wrath, my wrath did end.
I was angry with my foe:
I told it not, my wrath did grow.
(5) And I water'd it in fears,
Night & morning with my tears;
And I sunned it with smiles,
And with soft deceitful wiles.
And it grew both day and night,
(10) Till it bore an apple bright;
And my foe beheld it shine,
And he knew that it was mine,
And into my garden stole
When the night had veil'd the pole:
(15) In the morning glad I see
My foe outstretch'd beneath the tree.

31. What kills the speaker's enemy?

    (1) A pole

    (2) Water

    (3) A poison apple

32. A good synonym for "wrath" is

    (1) sadness.

    (2) anger.

    (3) deceit.

33. The speaker's wrath for his foe grew because

    (1) the speaker did not talk about it.

    (2) the speaker's foe made him angrier.

    (3) the speaker's foe wouldn't listen to him.

34. The speaker's foe died because

    (1) he continued to lie to the speaker.

    (2) he destroyed the speaker's garden.

    (3) he stole an apple from the speaker's garden.

35. The poison in this poem is

    (1) a metaphor for anger.

    (2) alliteration.

    (3) a paradox.

36. What is the theme of this poem?

    (1) When we are angry at someone we should talk to our friends about it.

    (2) When we don't talk things through with our enemies, our anger can become deadly.

    (3) When you steal, whether from an enemy or friend, you must suffer the consequences.

37. The rhyme scheme of this poem is

    (1) *abaa cdcc.*

(2) *abab cdcd.*

(3) *aabb ccdd.*

Questions 38–42 refer to the following passage.

PASSAGE H

The morning's brightness drew me awake and I was surrounded with strangeness. I had slid down the seat and slept the night through in an ungainly position.
(5) Wrestling with my body to assume an upward arrangement, I saw a collage of Negro, Mexican and white faces outside the windows. They were laughing and making the mouth gestures of talkers but their
(10) sounds didn't penetrate my refuge. There was so much curiosity evident in their features that I knew they wouldn't just go away before they knew who I was, so I opened the door, prepared to give them
(15) any story (even the truth) that would buy my peace.

The windows and my grogginess had distorted their features. I had thought they were adults and maybe citizens of
(20) Brobdingnag, at least. Standing outside, I found there was only one person taller than I, and that I was only a few years younger than any of them. I was asked my name, where I came from and what led me to the
(25) junkyard. They accepted my explanation that I was from San Francisco, that my name was Marguerite but that I was called Maya and I simply had no place to stay. With a generous gesture the tall boy, who
(30) said he was Bootsie, welcomed me, and said I could stay as long as I honored their rule: No two people of opposite sex slept together. In fact, unless it rained, everyone had his own private sleeping accommoda-
(35) tions. Since some of the cars leaked, bad weather forced a doubling up. There was no stealing, not for reasons of morality but because a crime would bring the police to the yard; and since everyone was under-
(40) age, there was the likelihood that they'd be sent off to foster homes or juvenile delinquent courts. Everyone worked at something. Most of the girls collected bottles and worked weekends in greasy spoons.
(45) The boys mowed lawns, swept out pool halls and ran errands for small Negro-owned stores. All money was held by
(25) Bootsie and used communally.

From *I Know Why the Caged Bird Sings*, by Maya Angelou (1973)

38. The author spent the night in

    (1) a hotel.

    (2) a junkyard.

    (3) San Francisco.

39. The atmosphere in this community can best be described as

    (1) leisurely.

    (2) every man for himself.

    (3) cooperative.

40. All of the people the author met

    (1) were homeless.

    (2) were thieves.

    (3) worked in the junkyard.

41. The age of everyone in this community was probably between

    (1) 1–9.

    (2) 10–17.

    (3) 18–25.

42. In the first paragraph, "ungainly" means:

    (1) fetal.

(2) comfortable.

(3) awkward.

Questions 43–45 refer to the following passage.

PASSAGE I

A true classic, as I should like to hear it defined, is an author who has enriched the human mind, increased its treasure and caused it to advance a step; who has dis-
(5) covered some moral and not equivocal truth, or revealed some eternal passion in that heart where all seemed known and discovered; who has expressed his thought, observation, or invention, in no matter
(10) what form, only provided it be broad and great, refined and sensible, sane and beautiful in itself; who has spoken to all in his own particular style, a style which is found to be also that of the whole world, a style
(15) new without neologism, new and old, easily contemporary with all time.

From "What Is a Classic?"
by Charles Augustin Sainte-Beuve (1850)

43. According to the author, a true classic must be

    (1) stylish.

    (2) new.

    (3) timeless.

44. The passage, "in his own peculiar style, a style which is found to be also that of the whole world, a style new without neologism, new and old, easily contemporary with all time," is

    (1) a simile.

    (2) a paradox.

    (3) a soliloquy.

45. The author's first qualification for a classic is that

    (1) it must be intellectually stimulating.

    (2) it must be stylish.

    (3) it must start a new trend.

INTERPRETING LITERATURE AND THE ARTS

ANSWER KEY

1. (1)	13. (2)	25. (3)	37. (3)
2. (2)	14. (3)	26. (1)	38. (2)
3. (3)	15. (3)	27. (3)	39. (3)
4. (3)	16. (3)	28. (2)	40. (1)
5. (3)	17. (1)	29. (3)	41. (2)
6. (2)	18. (2)	30. (3)	42. (3)
7. (1)	19. (1)	31. (3)	43. (3)
8. (3)	20. (1)	32. (2)	44. (2)
9. (2)	21. (2)	33. (1)	45. (1)
10. (1)	22. (3)	34. (3)	
11. (3)	23. (2)	35. (1)	
12. (1)	24. (2)	36. (2)	

PRE-TEST SELF-EVALUATION

Question Number	Subject Matter Tested	Section to Study (section, heading)
1.	main idea of paragraph	II, From Active Reading to Comprehension
2.	main idea of paragraph	II, From Active Reading to Comprehension
3.	inference	II, From Active Reading to Comprehension
4.	elements of non-fiction	III, Elements of Non-Fiction
5.	types of essays (rhetorical modes)	III, The Essay
6.	autobiography	III, Non-Fiction, Autobiography
7.	autobiography	III, Non-Fiction, Autobiography
8.	autobiography	III, Non-Fiction, Autobiography
9.	structure/style	III, Elements of Non-Fiction
10.	inference	II, From Active Reading to Comprehension
11.	understanding character through dialogue	V, Elements of Drama
12.	understanding character through dialogue	V, Elements of Drama
13.	understanding character through dialogue	V, Elements of Drama
14.	inference	II, From Active Reading to Comprehension
15.	irony	V, Elements of Drama
16.	tone	IV, Reading Poetry, Elements of Sense
17.	figurative language	III, Elements of Fiction and IV, Reading Poetry, Elements of Sense
18.	figurative language	III, Elements of Fiction and IV, Reading Poetry, Elements of Sense
19.	figurative language, diction	III, Elements of Fiction and IV, Reading Poetry, Elements of Sense
20.	poetic structure	IV, Reading Poetry, Poetic Forms
21.	theme	IV, Reading Poetry, Elements of Sense
22.	theme, tone	IV, Reading Poetry, Elements of Sense
23.	understanding character	III, Elements of Fiction
24.	style, inference	III, Elements of Fiction
25.	vocabulary in context	II, Active Reading
26.	understanding character through details	II, Active Reading
27.	inference	II, From Active Reading to Comprehension

II = Reading Literature   III = Reading Prose   IV = Reading Poetry   V = Reading Drama   VI = Reading Commentary

Pre-GED Interpreting Literature and the Arts

Question Number	Subject Matter Tested	Section to Study (section, heading)
28.	main idea of paragraph	II, From Active Reading to Comprehension
29.	realism	III, Elements of Fiction
30.	inference	II, From Active Reading to Comprehension
31.	plot in narrative poetry	IV, Reading Poetry, Poetic Forms
32.	vocabulary in context	II, Active Reading
33.	plot in narrative poetry	IV, Reading Poetry, Poetic Forms
34.	plot in narrative poetry	IV, Reading Poetry, Poetic Forms
35.	figurative language	IV, Reading Poetry, Elements of Sense
36.	theme	IV, Reading Poetry, Elements of Sense
37.	rhyme schemes	IV, Reading Poetry, Poetic Forms
38.	setting	III, Elements of Fiction
39.	inference	II, From Active Reading to Comprehension
40.	inference	II, From Active Reading to Comprehension
41.	identifying details	II, Active Reading
42.	vocabulary in context	II, Active Reading
43.	opinion in commentary	VI, Commentary
44.	paradox	IV, Reading Poetry, Elements of Sense
45.	opinion in commentary	VI, Commentary

II = Reading Literature   III = Reading Prose   IV = Reading Poetry   V = Reading Drama   VI = Reading Commentary

PRE-TEST ANSWERS AND EXPLANATIONS

Passage A

1. **(1)** The author went to the woods to (1), discover the meaning of life. There are several different clues that suggest this is the reason he went to the woods, primarily, there is the repetition of the word "life" in the first paragraph. In addition, the first sentence tells us that the author wanted "to front only the essential facts of life, and see if I could not learn what it [life] had to teach." He desired to "reduce it [life] to its lowest terms" and "to know it by experience" and then "give a true account of it." The other answers may have also been part of the reason he went into the woods, but they are not specifically referred to in the passage; he does not speak of going to (2), be alone, or (3), write. (He does speak of publishing his findings, but he goes to the woods to find out about life, not to write.)

2. **(2)** According to the author, what is wrong with our lives is that there are (2), too many details and distractions. This is evident from the second paragraph, where the author writes "Our life is frittered away by detail." He begs his readers to simplify their lives so that they can "keep [their] accounts on [their] thumb-nail[s]." There is no mention of honesty in the passage, so (1), not enough honest people, is not a correct answer. He does mention fighting ("like pygmies we fight with cranes"), but this is just one example of his larger point that "life is frittered away by detail."

3. **(3)** The italicized phrase "somewhat hastily" suggests that the author (3), believes that "to 'glorify God and enjoy him forever' " is *not* the chief end of life. The italics emphasize the author's belief that men have come to this conclusion too hastily. Also, he tells us that "most men…are in a strange uncertainty about it," so we can infer that he doubts their conclusion is correct.

4. **(3)** This passage is (3), non-fiction. It is not commentary, because it is not discussing a work of art; and because it is in prose form, not verse, we know it is not (1), poetry.

5. **(3)** The author's purpose in this passage is to (3), convince. This is evident from his repetition of "Simplicity, simplicity, simplicity!" and his appeal to his readers: "Yet your affairs be two or three, and not a hundred or a thousand; instead of a million count a half a dozen, and keep your accounts on your thumb-nail." (It is possible to say that the passage's purpose is (1), to explain why he went into the woods, but the author explains why he went to the woods so that readers would see the logic in it and do the same.)

Passage B

6. **(2)** Westley got up because (2), he got tired of waiting: "God damn! I'm tired o'sitting here. Let's get up and be saved." Answer (1), because his mother told him to, cannot be correct since there is no mention of

Westley's mother in the passage; (3), because Hughes got up, cannot be correct because Westley got up before Hughes did.

7. **(1)** The author was waiting (1), for Jesus to show himself. He tells us in paragraph two: "And I kept waiting serenely for Jesus, waiting, waiting—but he didn't come. I wanted to see him, but nothing happened to me." Thus answers (2) and (3) are incorrect.

8. **(3)** The author got up because (3), he was making everyone wait. The author tells us that "it was really getting late" and "I began to be ashamed of myself, holding everything up so long." He also "decided that maybe to save further trouble" for the others, he'd "better lie too" and get up. He did not get up because (1), he had to go to the bathroom (there is no indication of that in the passage); (2), he saw Jesus—we know that he didn't. It is clear that the author got up because he felt badly about making everyone wait so long.

9. **(2)** "So I got up" is in a paragraph by itself because (2), it was what everyone was waiting for. We know that the author was the last one to get up, and we know that the author felt badly because everyone was waiting for him. We know that the author did not (1), only imagine getting up or (3), finally see Jesus.

10. **(1)** We can conclude that the author was (1), misled about what would happen in church. This is suggested by the fact that he kept waiting to actually *see* Jesus (paragraph 2). He was the last to get up because he kept waiting for Jesus to appear. We can't conclude that the author was (2), very religious, because it is clear that he has not had much experience in church and does not know what it means to be "saved." Finally, we have no evidence in the passage that he was (3), very good friends with Westley. In fact, the way Westley is introduced—"He was a rounder's son named Westley"—indicates that they were merely acquaintances.

Passage C

11. **(3)** Biff is home because (3), he isn't getting anywhere in life. He tells his brother that "whenever spring comes to where I am, I suddenly get the feeling, my God, I'm not gettin' anywhere!…That's when I come running home." He also says that "every time I come back here I know that all I've done is to waste my life."

12. **(1)** Biff thinks Happy is happy because (1), he has money. Happy does mention that he has (2) "plenty of women," and that he has (3), an apartment and a car, but we can tell that (1) is correct because when Happy tells Biff he's not happy, Biff asks, "Why? You're making money, aren't you?"

13. **(2)** Happy can best be described by sentence (2), money can't buy you happiness. Even though Happy has money, he tells Biff that he is not happy (see explanation for question 12, above).

14. **(3)** The merchandise manager isn't happy because (3), the more he has, the more he wants. Happy tells us that the manager "just built a terrific estate on Long Island" but that he only lived there for two months because "He can't enjoy it once it's finished." He wants something more, so "he's building another one." We know that the manager has money, so (1) isn't the answer, and there is no indication in the passage that the merchandise manager knows that (2), Happy is after his job.

15. **(3)** We can conclude from this passage that Happy's name is (3), ironic. Happy, the passage reveals, is *not* happy, so his name is the opposite of its reality. This is irony.

Answers and Explanations

Passage D

16. **(3)** The adjective that best describes how the speaker feels about the divorce is (3), empty. The poem tells us that the couple feels like Siamese twins (line 7) that have now been "severed" (line 13). There is a part of each of them missing now.

17. **(1)** Line 4, "Drawn tight, it could choke us," suggests that (1), marriage can strangle love. The "it" in the line refers to the "garland" (line 1) of marriage. The garland, which was often worn for marriage ceremonies, represents the bond of marriage. As line 4 suggests, if that garland (their bond) is drawn too tight, it could strangle them (their love).

18. **(2)** "Siamese twins" in line 7 is (2), a metaphor. A metaphor is a comparison that does not use "like" or "as." Here, the speaker compares their closeness to the closeness of Siamese twins whose bodies are physically connected (and often inseparable, as they may share vital organs or other essential body parts).

19. **(1)** "Yoke" in line 6 suggests (1), marriage requires work. A yoke is the wooden harness fastened over the necks of oxen or other animals used for difficult physical labor. This suggests that their marriage required some difficult labor as well; it was not always easy.

20. **(1)** This poem is (1), free verse. Blank verse (2) has no rhyme scheme but has a metrical pattern; a sonnet (3) has a specific rhyme scheme and a metrical pattern. This poem has neither a rhyme scheme nor a metrical pattern.

21. **(2)** The sentence that best sums up the relationship between the speaker and spouse is (2), they had become like one person. This is best suggested by the Siamese twins metaphor as well as lines 8–13, which expresses their doubt that they will be able to survive now that they have been "cut apart" (line 10).

22. **(3)** The divorce may (3), kill them. Again, this is suggested by lines 8–13 in which they doubt their ability to survive without each other after their divorce. They're waiting to see "if we can survive, / severed" (lines 12–13).

Passage E

23. **(2)** The woman described in this passage is (2), a member of the working class. She cannot be (1), a member of the nobility, because we learn that she "had no dowry," that she "dressed plainly," and that she "suffered from the poverty of her dwelling." She cannot be (3), a servant, because we learn in the last paragraph that she has a servant of her own—a "little Breton peasant" who does her housework. We also know she married a clerk, and a clerk is a typical working-class position.

24. **(2)** The phrase "she let herself be married to a little clerk" suggests that (2), she thought she was above marrying him. This is evident from the first paragraph, where we learn that she had no hope of "being known, understood, loved, wedded by any rich and distinguished man." She "let herself be married to a little clerk" because she would not be able to do any better. There is no indication in the passage that she (1), let her husband do whatever he wanted, and the passage makes it clear that she (3), wanted to marry a rich man, not a clerk.

25. **(3)** The word "ceaselessly" in paragraph 3 means (3), constantly. We can tell that her suffering was constant because that paragraph lists item after item that made her suffer: the walls, the chairs, the curtains, the housekeeper that she had; the antechambers, candelabra, footmen that she didn't have. Thus, (1), occasionally, is not correct, and answer (2),

deservedly, is not quite logical in the context of the sentence, which informs us that she feels she didn't deserve to suffer so.

26. **(1)** She differed from other women of her rank because (1), she felt she deserved a higher rank. She didn't differ because (2), she married a clerk (she herself was from a family of clerks), and she "let herself" be married to him. She may have differed because she (3), had a taste for elegance, but we would have difficulty finding evidence for this in the passage. We can, however, find evidence for the fact that she thought she deserved a higher rank—"She suffered ceaselessly, feeling herself born for all the delicacies and all the luxuries"—and we can find evidence that other women of her rank didn't share this opinion: "All those things, of which another woman of her rank would never even have been conscious, tortured her and made her angry."

27. **(3)** She says rank (3), is determined by her beauty and charm, not her birth. This is expressed in paragraph two: "since with women there is neither caste nor rank: and beauty, grace, and charm act instead of family and birth. Natural fineness, instinct for what is elegant, suppleness of wit, are the sole hierarchy, and make from women of the people the equals of the very greatest ladies."

Passage F

28. **(2)** The writer found Maupassant's stories so moving because (2), they were written from a new perspective. She writes: "Here was a man who had escaped from tradition and authority, who had entered into himself and looked out upon life through his own being and with his own eyes, and who, in a direct and simple way, told us what he saw." We know the answer can't be (1), they were written just for her, because she did not know him personally (she merely "stumbled upon" his writing).

We know it can't be answer (3), it was fashionable to read them, because she writes that he had "gone out of fashion."

29. **(3)** "Here was life, not fiction" suggests that (3), Maupassant was a realist. He may have written biographies (1), or been a reporter (2), but the subject of this passage is Maupassant's stories, so we know that the author of this passage is referring to Maupassant's *fiction*. Fiction that is lifelike is realist fiction—stories that attempt to mimic reality, to be as true to life as possible.

30. **(3)** The writer says Maupassant "gives us the best that he can" because (3), he shares something of himself in his fiction. The writer of this passage tells us that Maupassant "entered into himself and looked out upon life through his own being and with his own eyes" and "told us what he saw"—he "gives us his impressions." The repetition of "his" in these excerpts emphasizes that the writer feels Maupassant shares himself.

Passage G

31. **(3)** It is (3), a poison apple, that kills the speaker's enemy. In line 10, the poem tells us that the speaker's wrath, which he watered and sunned (second stanza), grew "Till it bore an apple bright." In the last stanza, we learn that the enemy "stole" into the speaker's garden and died "beneath the tree." We can assume that this is the tree that bore the apple and that the apple was deadly.

32. **(2)** A good synonym for "wrath" is (2), anger. This is revealed in the first stanza, which alternates the two words: "I was angry with my friend: / I told my wrath, my wrath did end. / I was angry with my foe: / I told it not, my wrath did grow."

Answers and Explanations

33. **(1)** The speaker's wrath for his foe grew because (1), the speaker did not talk about it. This is expressed in line four: "I told it not, my wrath did grow." This is in contrast to line two, which reveals that the speaker's wrath for his friend subsided because he talked about it. Furthermore, there is no indication in the poem that (2), the speaker's foe made him angrier, or (3), the speaker's foe wouldn't listen to him.

34. **(3)** The speaker's foe died because (3), he stole an apple from the speaker's garden. See the explanation for question 31.

35. **(1)** The poison in this poem is (1), a metaphor for anger. The speaker's anger is compared to poison which killed the speaker's foe. Alliteration, (2), is the repetition of sounds; a paradox, (3), is a statement that seems to contradict itself but which contains some truth.

36. **(2)** The theme of this poem is (2), when we don't talk things through with our enemies our anger can become deadly. This is suggested by the action—and lack of action—taken by the speaker. In the first two lines, the speaker tells us he was angry with his friend, but spoke to him about it and the anger subsided. But he did not speak to his enemy about his anger. Instead, he kept it to himself, and his anger grew until it "bore an apple bright" whose poison killed his foe (the speaker found his foe "outstretch'd beneath the tree," and was glad about it).

37. **(3)** The rhyme scheme of this poem is (3), *aabb ccdd*. The lines are rhymed in pairs: friend, end; foe, grow; fears, tears; smiles, wiles; etc.

Passage H

38. **(2)** The author spent the night in (2), a junkyard. This is revealed in the second paragraph when the writer tells us what the strangers said to her: "I was asked my name, where I came from and what led me to the junkyard."

39. **(3)** The atmosphere in this community can best be described as (3), cooperative. We learn that "[e]veryone worked at something" and that "[a]ll the money was held by Bootsie and used communally." Thus, (1), leisurely, cannot be the correct answer (since everyone worked). Answer (2), every man for himself, cannot be correct, since everyone worked as a team and cooperated.

40. **(1)** All of the people the author met were (1), homeless. We learn that they do not steal, (so answer (2), thieves, cannot be correct). They don't steal because "a crime would bring the police to the yard; and since everyone was underage, there was the likelihood that they'd be sent off to foster homes or juvenile delinquent courts." If they would be shipped to foster homes or courts, we can assume that they either did not have homes or that they were run-aways. We also know that they lived in the junkyard (everyone had "his own private sleeping accommodations"—a car) and that they worked "in greasy spoons" and pool halls, so they did not (3), work in the junkyard.

41. **(2)** The age of everyone in this community was probably (2), between 10–17. We know that they are underage, so answer (3) 18–25, cannot be correct. We also know that they are old enough to work and that they are old enough that a rule forbidding people of the opposite sex to sleep together is necessary. Thus, (2) is the best answer.

42. **(3)** In the first paragraph, "ungainly" means (3), awkward. We know that the author "had slid down the seat and slept the night through in an ungainly position" so that she had to "wrestle" her body "to assume an up-

ward arrangement." Thus, we can assume that her position was not (2), comfortable. Since she slid down the seat and was on the floor of the car, it is unlikely that she had enough room to curl up (1), into a fetal position. Furthermore, we know she had to struggle to sit back up, so her position must have been awkward.

Passage I

43. **(3)** According to the author, a true classic must be (3), timeless. This can be inferred from the last part of the paragraph, in which the author states that a classic must be "new and old, easily contemporary with all time." If it is both new and old, it must be timeless. Answer (2), new, cannot be correct because the author writes that it must be new *and* old, not just new. Answer (1), stylish, cannot be correct either because if something is stylish, it is a common style or fashion, and the passage tells us that a classic must have its "own particular style." This question requires a very careful reading of the passage.

44. **(2)** The passage "in his own peculiar style, a style which is found to be also that of the whole world, a style new without neologism, new and old, easily contemporary with all time" is (2), a paradox. A paradox is an apparent contradiction that has an element of truth to it. This sentence pairs a series of opposites, like new and old. These pairings seem contradictory; after all, how can something be both old and new? Yet, a literary classic is just that; it speaks to all ages and eras. Thus, (2) is the correct answer. A simile, (1), is a comparison using "like" or "as"; a soliloquy, (3), is a lengthy speech by a character in a play in which that character reveals his or her thoughts as if he or she is alone and thinking aloud.

45. **(1)** The author's first qualification for a classic is (1), it must be intellectually stimulating. The first qualification listed in the paragraph "is an author who has enriched the human mind, increased its treasure and caused it to advance a step." An author who enriches the human mind and causes it to advance must stimulate thought.

Interpreting Literature
and the Arts

Reading Literature

INTERPRETING LITERATURE AND THE ARTS

READING LITERATURE

WHAT *IS* LITERATURE? (A WORKING DEFINITION)

Technically, **literature** can be defined as any written or published text, from a collection of poems to a shopping list. The type of literature we will discuss, however, and the type of literature you will be tested on in the GED, is generally defined as those writings we value not only for the *message they convey,* but also for their *beauty of form* and *emotional impact.*

By this definition, a short story would be considered literature, whereas a VCR user manual would not. We appreciate the manual for its information only, not for any beauty of form or emotional impact. Furthermore, the message conveyed in the user manual, unlike the kind of messages conveyed through literature, will not have an impact on our values or beliefs. Literature, on the other hand, often forces us to examine our perceptions, assumptions, values, and beliefs—of ourselves, others, and the world around us.

Within this definition, the type of texts that classify as literature are numerous. There are novels, short stories, poems, plays, biographies, autobiographies, reviews, and essays of all sorts. We will discuss each of these forms according to the following categories or **genres**: **prose, poetry, drama,** and **commentary**.

Typically, literature—both in its **fiction** (invented story) and **non-fiction** (true story) forms—asks us to come to a certain conclusion or judgment about the people, places, and/or events described. Sometimes, as in an **expository essay** (an essay whose purpose is to explain or demonstrate), the conclusion the author wants us to come to is clear from the outset. This idea comes across in the essay's **thesis** statement, which reveals the author's position and purpose. Other times, however, as in works of fiction, the author does not provide us with such a clear statement of position and purpose. If this is the case, we must work to discover the author's **theme**, the overall meaning or idea he or she wants to convey.

Thus, when reading literature, especially fiction, we often assume the role of detectives: as we read, we must look for clues to piece together the text's meaning. Even a well-written essay with a clear thesis, such as Richard Selzer's "The Knife," still requires the reader to search for meaning. Like all good literature, "The Knife" has several layers of meaning—several levels at which the essay can be read and interpreted. The deeper the level—the more careful the reading—the more significant our understanding. True, it is sometimes difficult and challenging to understand literature. But if we accept that challenge, we will also find that reading literature is immensely rewarding.

READING LITERATURE: A SEARCH FOR MEANING

Because literature involves more than simply conveying a message or set of instructions, understanding literature requires a certain degree of "training." This section will provide you with specific strategies to help you become a better reader of literature.

PRE-READING

There are a number of things you can do to increase your comprehension even *before* you read a particular passage. When you are given a text, first:

- read the title carefully;
- notice the name of the author;
- look for the date of publication;
- read the questions you'll be asked to answer about that text.

The title will usually suggest both the subject matter of the text and the author's attitude toward it. It will also set up certain expectations for the passage. For example, an essay titled "The End of the World Is Near" will probably deal with a very serious issue, such as nuclear weapons or global warming. An essay like Ann Lovejoy's "Friends of Dirt, Unite!" creates a totally different set of expectations. We can sense from the title alone that this will be a light-hearted essay about dirt. We can also tell that Lovejoy is "pro-dirt" and is reaching out to others who are "pro-dirt" like she.

What about the author? If it happens to be an author you've read before, then you already know something about the passage. You should have an idea of the author's style and perhaps even know something of his or her philosophies. If you've never read the author before, see what you can determine from the name. Is the author male or female? Can you identify the nationality? These distinctions may give you some indication of the author's background and beliefs. Do be careful, though: The author may have taken a married name or may be writing under an assumed name (**pseudonym**), so your assumptions based on gender and nationality are always just that—assumptions.

The date of a text's publication is also very revealing. When was it published? A story published in 1600, for example, will be very different from a story published in 1900. An essay published in 1960 will be very different from one published in 1990. Recall what you can about the time period in which the text was written. Consider the social, political, and moral climate of the world at that time.

Finally, when you read the questions that you will have to answer about that text, you give yourself an "awareness advantage." As you read, you will be able to keep your eye out, so to speak, for the answers you'll need later. By no means should you read for the answers alone, though. Often answers come only from an understanding of the text as a whole.

ACTIVE READING

The key to reading and understanding literature is **active reading**. Though we may seem passive when we read since we are usually sitting or lying still, our minds should be actively engaged with the text. With the possible exception of the most **didactic** of tales, like fables, where the "moral of the story" is obvious, we need to read literature as carefully as we would listen to the author were he or she speaking directly to us. All writing is, after all, communication. Whatever type of literature you are reading, remember that the writer is seeking to convey his or her ideas to you, the reader.

There are several elements of active reading, but we can group them into two basic steps:

- determining the meaning of unknown vocabulary words;
- marking up the text.

1. Determining the Meaning of Unknown Vocabulary Words

When we read, we often come across words that are unfamiliar to us. Whenever we read, then, we should have a dictionary handy, if possible, to look up any terms we don't know or aren't sure about. This may slow you down at first, especially if the text you're reading uses sophisticated or technical vocabulary. But if you get in the habit of stopping to look up words you don't know and reading the words in **context** (the surrounding words and ideas that suggest meaning), you will (1) improve your vocabulary and (2) help ensure a complete understanding of the text.

If you are wise, once you've looked up a new word, you will write its meaning down—preferably in the margin of the text you are reading, next to the word in question. By writing it down, you'll help seal the meaning of the word in your memory. By writing it in the margin, you'll ensure that the meaning is there for you next time you return to the text. If you are unable to write in the margins, keep a vocabulary list on a separate sheet of paper or in a notebook.

If you don't have a dictionary available when you're reading, make sure you circle the unfamiliar words and/or write them down to look them up later. And do make sure you look them up. Then, after you've found the meaning of each word, re-read the section of the text where each word is used. This way you can review the use of each word in context and see how each word affects the meaning of the text.

But what if you can't use a dictionary, like during an exam? Does that mean you must forfeit meaning? No, not at all. Even if you don't have a dictionary available, you can generally make an educated guess as to the meaning of a word based on its **context**—the way the word is used in the sentence, how it fits in with the words and ideas around it. For example, in the following passage, see if you can determine what timorous means from the context provided by the sentence:

> She slowly poked her head around the corner, timorous as a turtle afraid to come out of its shell.

Timorous means

(a) bold

(b) happy

(c) sad

(d) worried

(e) shy

The correct answer is (e), shy. How can you tell? The words *slowly* and *afraid* in the sentence suggest this meaning. Because of these words, timorous could not mean bold or happy. Sad and worried are more likely possibilities, but shy is best suggested by "afraid to come out of its shell," and a shy person would be likely to move slowly and cautiously. Also, if you know the word timid—which means not bold, shy—you can see that *tim*orous shares the same root.

Now try another. Determine what accord means in the following sentence:

> Finally, after hours of debate and disagreement, the committee reached an accord, and a new policy was established.

An accord is

(a) a crossroads

(b) a roadblock

(c) an agreement

(d) a challenge

(e) a reversal

You can see that an accord takes place after *debate* and *disagreement*, and the result of it is "a new policy" agreed upon by the committee. Accord, therefore, means (c), an agreement.

Now try one more. Determine the meanings of succumb and steadfast in the following sentence:

> He thought I would succumb to his wishes if he flattered me enough, but I steadfastly refused to do what he wanted.

To succumb means

(a) to refuse

(b) to respect

(c) to share

(d) to give in

(e) to respond

Steadfastly means

(a) quickly

(b) firmly

(c) stupidly

(d) slowly

(e) selfishly

You can see that succumb means (d), to give in, because "he" wants something from "me," and he is going to try to convince me to do it by flattering me. It can't mean (a), because he wouldn't want me to refuse, and it can't mean (e), because wishes require more than a response. Choices (b) and (c) are also not possible, because "he" wants "me" to do something, and that requires more than respect or sharing. Steadfastly means (b), firmly. This should be clear because he is going to try to "flatter me enough," so he will be persistent and will require a strong, steady refusal.

You can also determine word meanings by looking at their prefixes, suffixes, and roots. For a list of common prefixes, suffixes, and word roots, see **Appendix B** at the back of this book.

2. **Marking Up the Text**

The second task to undertake as an active reader is to mark up the text as you read. This includes four specific strategies:

- underlining key words and ideas;

- recording your reactions in the margins (for example, "I like the way he says this," or "I totally disagree");

- recording your questions in the margins ("Why does the author use this term to describe his mother?" "Why does she tell her story out of chronological order?");

- noticing as much as you can about how the text is actually written (making observations).

It is this last step we will discuss next, for this strategy can help you find those key ideas and answer those questions you write in the margin.

FROM ACTIVE READING TO COMPREHENSION: MAKING OBSERVATIONS AND DRAWING INFERENCES

After your first reading, in which you should employ the first three strategies above, you should read the passage again, this time more slowly and carefully, to employ the last strategy: **making observations**. It may surprise you, but this first step to comprehension actually has little to do with interpretation. Before you try to decipher a text, you should first simply *look* at it—look closely at how the text is written. Observe the words the author uses, the style of the writing, the structure of the sentences and paragraphs, and the punctuation used. All writers use form and language to convey their meaning, so *how* they say what they say will help us understand what they *mean* by what they say.

Below is an excerpt from Richard Selzer's essay "The Pen and the Scalpel." It has been marked up to show you just how these active reading strategies might be employed.

Now, take a look at the following paragraph from Dick Gregory's narrative essay "Shame." Read it carefully, marking it up as you go. Try to determine the meaning of any unfamiliar words in context, and then look them up in the dictionary to check yourself. Then read the passage a second time and see how many things you can notice *about how it's written*. (Be sure you stick to concrete observations about the writing rather than general statements about the action or meaning of the paragraph.) For example: "I notice the first two lines have the words *everywhere* and *everyone*." Try to list at least five observations. Some words and phrases are italicized to help you with some initial observations.

Now *there was shame everywhere*. It seemed like the whole world had been inside that classroom, everyone had heard what the teacher had said, everyone had turned around and felt sorry for me. *There was shame* in going to the Worthy Boy's Annual Christmas Dinner for *you and your*

---

of the sky; of heaven, divine

Similarities:
1. both are "celestial arts"
2. both slender instruments leave a trail
3. both instruments are restrained
4. both "sew"
5. both have beginning, middle, and end.

At first glance, it would appear that surgery and writing have little in common, but I think that is not so. For one thing, they are both sub-(celestial) arts; as far as I know, the angels (disdain) to perform either one. In each of them you hold a slender instrument that leaves a trail wherever it is applied. In one, there is the shedding of blood; in the other it is ink that is spilled upon a page. In one, the scalpel is restrained; in the other, the pen is given rein. The surgeon sutures together the tissues of the body to make whole what is sick or injured; the writer sews words into sentences to fashion a new version of human experience. A surgical operation is rather like a short story. You make the incision, (rummage) around inside for a bit, then stitch it up. It has a beginning, a middle and an end. If I were to choose a medical specialist to write a novel, it would be a psychiatrist. They tend to go on and on. And on.

—What's the "invasion" in writing?

to regard worthy of contempt or scorn; to refrain because of disdain

I never thought about it this way

— why this word?

writing and surgery really <u>do</u> have a lot in common!

— does he not like novels?

kind, because everybody knew what a worthy boy was. *Why* couldn't they just call it the Boy's Annual Dinner, *why*'d they have to give it a name*? There was shame* in wearing the brown and orange and white plaid mackinaw the welfare gave to three thousand boys. *Why*'d it have to be the same for everybody so when you walked down the street the people could see you were on relief? It was a nice warm mackinaw and it had a hood, and my Momma beat me and called me a little rat when she found out I stuffed it in the bottom of a pail full of garbage way over on Cottage Street. *There was shame* in running over to Mister Ben's at the end of the day and asking for his rotten peaches, *there was shame* in asking Mrs. Simmons for a spoonful of sugar, *there was shame* in running out to meet the relief truck. I hated that truck, full of food for *you and your kind*. I ran into the house and hid when it came. And then I started to sneak through alleys, to take the long way home so the people going into White's Eat Shop wouldn't see me. Yeah, the whole world heard the teacher that day, we all know you don't have a Daddy.

List your observations here. Remember that there are no right or wrong answers; simply look and write down what you notice.

I noticed:

1. _____
2. _____
3. _____
4. _____
5. _____
6. _____

7. _____

Did you look up *mackinaw*? If you didn't, and you don't know what it means, can you determine its meaning from the context? Check your answer in a dictionary.

Did you notice:

- How many times he uses the word "shame" in this paragraph?
- How many different examples of things that brought him shame he includes in this paragraph?
- That he repeats the phrase "you and your kind" several times?
- That in the eighth sentence, he lists several examples at once, rather than sentence by sentence?
- That he uses the pronoun "you" (second person point of view)?
- That he uses casual, conversational terms like "Yeah"?
- That he uses **alliteration**, the repetition of initial sounds, in line 28 with "*full of food for*"?
- The detail with which he describes the mackinaw and where he buried it?
- That he mentions "*my* Momma" but "*a* Daddy"?
- That he asks several questions in this paragraph, all beginning with "why"?
- That the first sentence makes an **assertion** (a statement or declaration that needs evidence or explanation to be accepted as true) about shame?
- That the rest of the paragraph serves to support that assertion (to "prove" that it's true) by listing examples ("evidence") of specific places where he felt shame?

How many of these observations did you make? If you didn't notice these things at first, reread the text. Can you see them now?

So, we've noticed all of these things about how Gregory wrote this paragraph from "Shame." Now what?

The next crucial step is to ask yourself *why*. *Why* does Gregory use the phrase "there was shame" so many times in this paragraph? *Why* does he use the pronoun "you" instead of "me" ("you and your kind" instead of "me and my kind")? *Why* does he include so many examples? Your task, in other words, is to move from observation to **inference**.

An **inference** is a conclusion based upon reason, fact, or evidence. Students often misinterpret literature because they make inferences based upon their own ideas or hunches. But if your inferences about literature are based on solid "facts"—in this case, your concrete observations about the writing—it is likely that your inferences will be correct and guide you to a more complete understanding of the author's ideas.

For example, look over the observations we've made about this paragraph from "Shame," and answer the following question:

Which statement best sums up the **main idea** of this paragraph?

(1) Gregory felt shame.

(2) Gregory was embarrassed about what the teacher said.

(3) Gregory could not escape his shame.

(4) Gregory was poor.

(5) Gregory was disliked by his teacher.

Although all of the above answers are true, the best answer is number 3: Gregory could not escape his shame. Before we examine why this is the best answer, let us talk briefly about the question. The question asks which statement "best sums up the **main idea**" of the paragraph. The main idea is different from the **subject** or **topic** of a text. The subject or topic is *who or what the text is about*. The main idea, on the other hand, tells us *how the author or character feels* about that subject.

The main idea of a paragraph is often expressed in a **topic sentence**. This sentence is usually—but certainly not always—found at the beginning of a paragraph in a sentence that makes an assertion about the subject of the paragraph. The rest of the sentences in the paragraph provide details or specific examples to support that assertion. When the main idea of a paragraph is found at the end of a sentence, that sentence will often begin with phrases or words like *in short, in brief, in fact, clearly, thus*, and *as these examples show*.

In narrative and other expository essays like "Shame," the main idea of the essay (the **thesis**) is often found in a **thesis statement**. This statement is usually found near the end of the first or second paragraph, depending upon the length of the essay. Like a topic sentence, it makes an assertion that must be supported by the paragraphs that follow.

Sometimes the main idea (of a paragraph or full text) will be left unsaid, and it is up to you to *infer* the writer's main idea. To help you determine the main idea in these cases, look carefully at the title of the text, which often indicates the author's attitude towards his or her subject. Look also at the examples and details the writer has provided and pay particular attention to the writer's choice of words. If the author describes a person as *quiet*, he or she expresses a certain attitude; *reserved*, another; *closed*, another still. These words essentially mean the same thing, but they convey different impressions.

Notice also the structure and tone of the paragraph. Then ask yourself, "What do all these things add up to? What do the examples the writer provides have in common? What idea would sum up all of these examples?" A main idea will make a general statement about all of the examples; it will encompass the whole paragraph or essay, not just a part of it. This is an important distinction to keep in mind. Very often readers will mistake an example for the main idea. If the idea doesn't cover <u>all</u> of the examples in that paragraph or essay, it is not the main idea.

We've said that the statement that best sums up the main idea of this paragraph from "Shame" is number 3: Gregory could not escape his shame. Why? All our observations point to this answer. First, look at the opening sentence of the paragraph: *"Now there was shame everywhere."* Gregory not only reveals the subject of this paragraph in this sentence; he also reveals how he feels about it. This is a very clear **topic sentence**.

Furthermore, as in all good texts, Gregory's writing reflects the idea he wishes to convey. Just as there was shame in his life, shame in everything around him and all that he did, there is also "shame" everywhere in this paragraph. Just as he could not escape his shame, the reader cannot escape the word "shame" in this paragraph. As readers, we feel like Gregory: overwhelmed by shame, burdened by it. We almost even begin to feel shame ourselves, especially because he uses the pronoun "you" ("you and your kind," "you don't have a Daddy") to make us feel as if the shame is indeed ours. This is also why he lists so many examples of places and things that made him feel shame. Had he just listed one or two, the paragraph would not have been as effective. But with example after example, we cannot deny that Gregory's shame truly was "everywhere."

Now, given the observations above and the inference you've just made, try another question.

Which sentence best describes Gregory's attitude about shame?

(1) Shame is a mild emotion and you shouldn't let it upset you.

(2) Shame can really keep you down.

(3) Shame is something to be ashamed of.

(4) Shame is a very powerful emotion that can cause you to behave irrationally.

(5) Shame is an illusion.

The best answer is number 4. Again, from the list of examples Gregory provides, we can see that his whole world was affected by shame, and this led him to some rather irrational behaviors, such as throwing away the warm, useful mackinaw and hiding when the relief truck came.

Making observations, then, is the first step to successful comprehension of literature. When we make observations about a text, some of the things we can notice are:

- important details about people, places, and things;
- repetition of words or phrases;
- specific words chosen to describe or explain;
- unusual uses of words or phrases;
- sounds of words (alliteration, etc.);
- sentence structure (statements vs. questions, long vs. short sentences, etc.);
- overall structure (division into sections, order of events or ideas, etc.).

The inferences we make based on our observations can help us:

- find the main idea;
- identify conflicts;

- predict outcomes;
- determine author's meaning and purpose;
- recognize author's strategies.

Now, let's practice. Take a careful look at Langston Hughes' poem "Mother to Son." Read the poem through twice, and then list your observations below.

MOTHER TO SON

(1) Well, son, I'll tell you:
(2) Life for me ain't been no crystal stair.
(3) It's had tacks in it,
(4) And splinters,
(5) And boards torn up,
(6) And places with no carpet on the floor—
(7) Bare.
(8) But all the time
(9) I'se been a-climbin' on,
(10) And reachin' landin's,
(11) And turnin' corners,
(12) And sometimes goin' in the dark
(13) Where there ain't been no light.
(14) So boy, don't you turn back.
(15) Don't you set down on the steps
(16) 'Cause you finds it's kinder hard.
(17) Don't you fall now—
(18) For I'se still goin', honey,
(19) I'se still climbin',
(20) And life for me ain't been no crystal stair.

I noticed:

1. _____
2. _____
3. _____
4. _____
5. _____
6. _____
7. _____

How much did you notice? Did you notice, for example, that:

- there is repetition of the phrase "I'se still"?
- there is repetition of the phrase "life for me ain't been no crystal stair," at the beginning and end?
- the mother speaks in **dialect** (language representing natural speech)?
- there are a lot of active, positive verbs, like climbing, turning, and reaching?
- the poem seems to turn on the word "so" in line 14?
- the mother makes a direct comparison between life and a staircase (a **metaphor**)?
- there's only one one-word line ("bare")?
- there's one speaker here (the mother)?
- there are a lot of apostrophes at the end of words instead of "g"?
- there are two groups of lines beginning with "and"?

Questions

Now, based on your observations, answer the following questions about "Mother to Son":

1. What is the mother's main message to her son?

    (1) Don't trip up.

    (2) Don't run away.

    (3) Don't count your chickens before they've hatched.

    (4) Don't give up.

    (5) Stay in school.

2. Why shouldn't the son quit?

   (1) She'll be ashamed of him if he does.

   (2) Things aren't that bad for him.

   (3) She is still climbing despite her obstacles.

   (4) There aren't many obstacles.

   (5) He'll never get a job if he does.

3. What is the mother's metaphor for life?

   (1) It is a dark room.

   (2) It is a staircase.

   (3) It is a tack.

   (4) It is honey.

   (5) It is crystal.

4. What kind of life has "mother" lived?

   (1) She has had an easy life.

   (2) She has been very busy raising children.

   (3) She has always found time to do things.

   (4) She has never had trouble getting what she needs.

   (5) She has faced a great deal of adversity.

Answers

1. **(4)** The message is (4), don't give up. We know this because the mother has continued to climb ("I'se still climbing") and has not given up ("I'se still goin' ") despite her obstacles (tacks, splinters). Also, she says "don't you turn back" (line 14) and "Don't you set down on the steps/ 'Cause you find it's kinder hard./ Don't you fall now" in lines 15–17.

2. **(3)** He shouldn't quit because (3), she has had many obstacles and she's still climbing. Lines 2–13 list the obstacles she's faced and describe her determination to continue, even when "there ain't been no light."

3. **(2)** Her life is (2), a staircase. She repeats the metaphor in lines 2 and 20 and uses the words "climbin' " and "landin's" to reinforce the metaphor.

4. **(5)** We can tell she has (5) faced a great deal of adversity, because of the condition of the stairs she's climbed. They weren't made of crystal; instead, they had tacks, splinters, and torn carpeting, if any. We know, therefore, the answer can't be (1), (3), or (4), and this is reinforced by her dialect—the way she speaks indicates she has not had much formal education. And we can't infer the answer is (2), because as far as we know from the poem, she only has one child. (Remember—even though it's likely that she has other children, the poem offers no evidence of this, so we cannot assume that she does; (2) is therefore not a valid answer.)

Practice: Applying Comprehension Strategies

DIRECTIONS: Read the following three passages. Apply the four reading strategies you've learned as well as your vocabulary skills. Write in the margins and mark up the text as you go. Then answer the questions following each passage.

PASSAGE A

The following excerpt from Charles Dickens' *Hard Times* introduces the character Thomas Gradgrind, a teacher. Read it actively (don't forget to write down your observations) and then answer the questions that follow.

Thomas Gradgrind, sir. A man of realities. A man of facts and calculations. A man who proceeds upon the principle that two and two are four, and nothing over,
(5) and who is not to be talked into allowing for anything over. Thomas Gradgrind, sir—peremptorily Thomas—Thomas Gradgrind. With a rule and a pair of scales, and the multiplication table always in his
(10) pocket, sir, ready to weigh and measure any parcel of human nature, and tell you exactly what it comes to. It is a mere question of figures, a case of simple arithmetic. You might hope to get some other nonsen-
(15) sical belief into the head of George Gradgrind, or Augustus Gradgrind, or John Gradgrind, or Joseph Gradgrind (all suppositions, non-existent persons), but into the head of Thomas Gradgrind—no, sir!
(20) In such terms Mr. Gradgrind always mentally introduced himself, whether to his private circle of acquaintance, or to the public in general. In such terms, no doubt, substituting the words "boys and girls," for
(25) "sir," Thomas Gradgrind now presented Thomas Gradgrind to the little pitchers before him, who were to be filled so full of facts.

1. What kind of man is Thomas Gradgrind? How can you tell? Use your observations to support your answer.

2. Would Gradgrind feel comfortable in a messy room? Why or why not? How can you tell?

3. What tools does Gradgrind use to explain human behavior?

4. What is Gradgrind's metaphor for his students? How does he plan to teach them?

5. Would you like to be one of Mr. Gradgrind's students? Why or why not?

PASSAGE B

The following excerpt is from "If Black English Isn't a Language, Then Tell Me, What Is?" by James Baldwin. Read it actively and then answer the questions that follow.

It goes without saying, then, that language is also a political instrument, means, and proof of power. It is the most vivid and crucial key to identity: It reveals the pri-
(5) vate identity, and connects one with, or divorces one from, the larger, public, or communal identity. There have been, and are, times, and places, when to speak a certain language could be dangerous, even
(10) fatal. Or, one may speak the same language, but in such a way that one's antecedents are revealed, or (one hopes) hidden. This is true in France, and is absolutely true in England: The range (and
(15) reign) of accents on that damp little island

make England coherent for the English and totally incomprehensible for everyone else. To open your mouth in England is (if I may use black English) to "put your busi-
(20) ness in the street": You have confessed your parents, your youth, your school, your salary, your self-esteem, and, alas, your future.

1. According to Baldwin, what does our language reveal about us? (Put your answer in your own words.)

2. True or false: According to Baldwin, the way that we speak can get us killed.

3. According to Baldwin, how is language our "most vivid and crucial key to identity"?

4. Do you agree with Baldwin that language reveals so much about us? Why or why not?

PASSAGE C

The following section, from *The House on Mango Street* by Sandra Cisneros, "Those Who Don't," is narrated by the main character, Esperanza. Read it carefully and then answer the questions below.

Those who don't know any better come into our neighborhood scared. They think we're dangerous. They think we will attack them with shiny knives. They are
(5) stupid people who are lost and got here by mistake.

But we aren't afraid. We know the guy with the crooked eye is Davey the Baby's brother, and the tall one next to him
(10) in the straw brim, that's Rosa's Eddie V. and the big one that looks like a dumb grown man, he's Fat Boy, though he's not fat anymore nor a boy.

All brown all around, we are safe. But
(15) watch us drive into a neighborhood of another color and our knees go shakity-shake and our car windows get rolled up tight and our eyes look straight. Yeah. That's how it goes and goes.

1. Why are the people who come to Esperanza's neighborhood scared?

2. Why shouldn't they be scared?

3. Why isn't Esperanza scared in her neighborhood?

4. Why is she scared when she goes to other neighborhoods?

5. What does Esperanza mean by the last sentence? Why do you think so?

Answers

Passage A

1. Thomas Gradgrind is a man who likes to measure, be precise, live according to formulas and scales. We can tell this because he is described as "a man of facts and calculations," a man "who proceeds upon the principle that two and two are four, and nothing more." He carries "a rule and a pair of scales" and a multiplication table to "weigh and measure any parcel of human nature." He reduces people to numbers and their feelings and actions to formulas and equations. He appears to believe he is superior to others because of his factual way of thinking (this is evidenced by the sentence "You might hope to get some other nonsensical belief into the head of George Gradgrind . . . but into the head of Thomas Gradgrind . . . no, sir!").

2. No, Thomas Gradgrind would not feel comfortable in a messy room. We can tell this because he is a very exact, precise, calculating man, and we can infer that he would be very upset if things were not precisely where they should be.

3. Thomas Gradgrind uses "a rule and a pair of scales" and "the multiplication table"—in other words, "figures" and "simple arithmetic"—to measure human nature. To Thomas Gradgrind, all human behavior boils down to formulas. This implies that people are mechanical, predictable, understandable, and controllable.

4. Thomas Gradgrind compares his students to "little pitchers," open vessels that he plans to "teach" by filling them up with "facts." In other words, he sees their minds as empty and his job as filling that void with facts—not with ideas or opinions, but facts. We can infer that Gradgrind does not want his students to think for themselves or to ask questions. Rather, he is the kind of teacher that wants students to repeat facts.

5. Answers to this question will vary. However, we can tell the author of this passage would not like to be one of Thomas Gradgrind's students. We see from the exaggerated importance Gradgrind places on formulas and facts that he is not a very *human* person; in fact, he does not even consider his students as people but as pitchers, useful tools and nothing more. We can infer that the author would prefer a teacher who treated the students as individuals with unique questions and concerns that cannot simply be weighed and measured according to formulas and scales.

Passage B

1. According to Baldwin, our language reveals who we are: what part of the world we are from, our ethnicity and ancestry, our education, our age, our social and economic status, our political involvement, and our self-image.

2. True: "There have been, and are, times, and places, when to speak a certain language could be dangerous, even fatal."

3. Language is our "most vivid and crucial key to identity" because language is what joins us to or separates us from other people. If we share a common language, then we are part of the larger community. If we don't, we are isolated from that community and unable to express who we are and what we feel. Also, language ties us to a larger cultural and historical identity.

4. Baldwin writes that the English use language to reveal a person's life (lines 10-20), but does not explain how they accomplish this feat. He writes that the French also use language to

reveal a person, but to a lesser degree than the English (lines 13 and 14). Because Baldwin offers no explanation, or proof, that language can reveal so much about us, the answer to this question must be based upon your personal experience.

Passage C

1. The people who come to Esperanza's neighborhood are scared because they think Esperanza's neighborhood is dangerous, that they will be attacked "with shiny knives."

2. They should not be scared because the people in Esperanza's neighborhood are not dangerous. They have simply been stereotyped as dangerous because they are "brown," and therefore different.

3. Esperanza isn't scared because she knows better; she knows the people in her neighborhood and knows they are not dangerous. Also, she knows everyone around her is "brown" like her, so she feels safe.

4. She is scared when she goes into a neighborhood "of another color" because she is afraid she will be attacked because she is "brown." She is, in other words, scared for the same reason that "those who don't know any better" are scared in her neighborhood: because she's different.

5. By "That's how it goes and goes," Esperanza is suggesting that this type of behavior has been around for a long time and will be around for a long time to come. People have always been prejudiced against people who look "different," and most people are afraid of what is unfamiliar to them. Furthermore, even those who experience this prejudice against them—people like Esperanza—still do the same thing in other neighborhoods. Esperanza is suggesting that this behavior is, unfortunately, a permanent part of human nature.

REVIEW

Literature is writing that is valued for its message, beauty of form, and emotional impact. A work of literature will ask us to come to certain conclusions about the people, places, and events it describes. That conclusion is driven by the author's thesis (in non-fiction and commentary) or theme (in fiction, poetry, and drama). Our response to the text and its themes will vary depending upon the experiences and ideas we bring to the reading.

Before we read literature we should pre-read by noting the title, author, and publication date of the text. If we are given a text in a test situation, we should read the questions before reading the text, so we can keep an eye out for the answers as we read.

When we read literature, we should read actively. Active reading involves two steps:

- determining the meaning of unknown vocabulary words;
- marking up the text.

If we don't have a dictionary, we can often determine the meaning of a word or phrase from the context in which that word or phrase is used. When we mark up a text, we should:

- underline key words and ideas;
- record our reactions in the margins;
- record our questions in the margins;
- make observations about the writing.

Our observations should be concrete things we've noticed about how the text is written (for example, how it is organized). The next step to comprehension is to make inferences based on what we've noticed and read. All conclusions we come to about literature should be based on something actually in the text itself, not just on our reactions. Therefore, it is important to find "evidence" (observations) that support the inferences we make.

The main idea of a paragraph is often expressed in a topic sentence, which makes an assertion about the subject of that paragraph. The main idea of an essay, its thesis, is often expressed in a thesis statement. This statement makes an assertion about the subject of the essay and is usually found near the beginning of the text. When the main idea is not clear, as it often isn't in fiction, poetry, or drama (which don't provide thesis statements), then we must look for clues as to the author's thesis or theme. These clues include the title, word choice, order of ideas, and examples in the text.

Interpreting Literature
and the Arts

Reading Prose

INTERPRETING LITERATURE AND THE ARTS

READING PROSE

DEFINING PROSE: FACT AND FICTION

What is **prose**? Essentially, prose is writing that is not in poetic form (verse) or dramatic form (stage or screenplay). There are two types of prose: **fiction** and **non-fiction**.

Fiction is prose that is the product of an author's imagination. What the author writes may be based in reality and may resemble actual people and events, but the characters and their experiences have been *invented* by the author. Fictional prose includes **novels** and **short stories,** as well as **myths, parables,** and **fables**.

Non-fiction, on the other hand, is prose that is <u>not</u> a product of the author's imagination. The people and events related in non-fictional prose are real. Non-fictional texts that we will discuss include **essays, biographies, autobiographies,** and **journalism** (magazine and newspaper articles).

A general distinction we can make between non-fiction and fiction is the same distinction we make between <u>fact</u> (non-fiction) and <u>fantasy</u> (fiction). Many authors today, however, are blurring that line by blending the two genres. Examples of such works are "autobiographical novels" such as Audrey Lorde's *Zami: A New Spelling of My Name* (a "biomythography") and Frederick Exley's *A Fan's Notes* (a "fictional memoir").

FICTION

Fiction comes from the fantastic realm of the imagination. These made-up stories often

help us make sense of the true stories that make up our lives. The word fiction comes from the Latin word *fingere*, which means "to make or shape." Fiction, then, is a story made or shaped by the author's imagination.

Human beings are natural storytellers. We tell stories for a number of reasons: to teach a lesson, to illustrate a point, to immortalize a person or event, to entertain, and to help ourselves and others understand the world around us.

Fiction appeals to us largely because of its structure: In all stories, there is a clear beginning, middle, and end. Though we may not be happy with a story's ending, we know it will conclude somehow, and we are anxious to find out which way the story will turn.

Unlike life, the events in a story have clear bounds; things start, complicate, and are usually resolved in some way by the end of the story. Given the lack of structure and lack of closure we often feel in our lives, it is no wonder we often turn to fiction: Stories present themselves as an ordered world in which the events that take place and the decisions made by the characters add up to something and make some sort of sense.

ELEMENTS OF FICTION

To better understand fiction, you should become familiar with these eight elements of fiction:

- plot
- setting
- character
- point of view
- tone
- language/style
- symbolism
- theme

Plot is the organization of events in a story—the order in which the action takes place. The plot is designed to build **conflict**, which is central to any good story. Without conflict of some kind, there is nothing to drive the action.

Stories are often, but certainly not always, plotted in chronological order. The author may vary the order of events to create more **suspense** or for other effects. Whatever the order of events, plots usually follow a five-part "pyramid" pattern of development. They begin with (1) **exposition**, an introduction to the people, places, and basic circumstances of the story. Authors often employ **foreshadowing**, the suggestion of things to come, in this section. Then the plot is (2) **complicated** by events that build up to the (3) **climax** of the story (the peak of the pyramid). The climax is the main **turning point** or moment of highest tension in a story. At the climax of a story, a character usually must make a difficult decision or take some kind of action.

After the climax, the missing pieces of the story are filled in (secrets are revealed, mysteries solved, and so on) and things fall into place in the (4) **falling action**. Finally, we read the (5) **resolution** or **denouement**, which "concludes" the story. The conflicts that drove the story are (to some degree, anyway) resolved, the questions answered, and lives straightened out.

Characters are those people created by the author to carry the action, language, and ideas of a story. As readers, it is our job to read about what these characters think, do, and say, and then try to understand *why* they think, do, and say such things.

Characters can be **round** or **flat**—that is, they can be three-dimensional, dynamic characters whose motivations we can see and understand, or they can be one-dimensional, undeveloped, static characters who we really know little about. Often, flat characters are **stereotypical** or **symbolic**.

In every story there is a **protagonist** and an **antagonist**. The protagonist is the "hero" or main character of the story, the person who faces the central conflict. The antagonist is the person (e.g., a villain), force (e.g., a war), or idea (e.g., prejudice) working against the protagonist.

Characters often reveal much about themselves through **dialogue**, the words they exchange with each other. Through dialogue, we learn what characters think, feel, and believe. We also learn something about their background and preferences. It is very important, therefore, to pay attention to the language a character uses. A character's dialect, for example—his or her natural speech represented through language—can help us determine where a character is from and therefore what experiences or beliefs he or she might have. For example, a character who says "Y'all come inside now" reveals that she is from the South, or that the story takes place in the South. Likewise, a character who says "Vat is dat?" tells us that he is from another country, perhaps Germany, and still has a heavy foreign accent.

While **dialect** reveals something about the character, it also reveals something about the author as well. Take, for example, the watchman in *The Adventures of Huckleberry Finn.* He says to Huck, 'Oh, dang it now, *don't* take on so; we all has to have our troubles, and this 'n 'll come out all right." Through the watchman's speech we learn that he has probably had a limited education; he says "we all has" instead of the grammatically correct "we all have." We also see that the author, Mark Twain, wants to represent his character as naturally as possible. By using dialect, he establishes a certain authenticity.

The **setting** of a story puts the characters and events in a particular time and place. It gives the story a specific social and historical context, thereby creating certain expectations in its readers. For example, if a story is set in Mississippi during the Civil War, we will have certain expectations: that there is political and social upheaval because of the war; that the majority of the characters are fighting to keep slavery; that any young white men in the novel will be fighting or on their way to fight; that women are still second-class citizens who do not yet have the right to vote; etc. These expectations may or may not be met in the story, but they are legitimate expectations based on the setting. (If they are not met, it is important to consider why the author has chosen to surprise us.)

Setting can also help create the **tone** of a story. The tone is the mood or attitude conveyed by the writing. For example, look at the first sentence of Edgar Allan Poe's short story, "The Fall of the House of Usher":

> "During the whole of a dull, dark, and soundless day in the autumn of the year, when the clouds hung oppressively low in the heavens, I had been passing alone, on horseback, through a singularly dreary tract of country; and at length found myself, as the shades of evening drew on, within view of the melancholy House of Usher."

The words *dull, dark, soundless, oppressively, alone, dreary,* and *melancholy* all serve to set a rather depressing, somewhat mysterious, tone to the story. In addition, it is significant that this sentence sets the story in autumn, the season when things begin to fall apart, so to speak, and the natural world begins to "die." This, then, also helps establish the proper tone for a horror tale. The tone of a story can, of course, change as the story progresses, but it is always linked to the action and the meaning of the story.

Perhaps the most important tone to consider in literature is **irony**. Irony in fiction and drama is when the reader knows more than one or more of the characters. The meaning of the character's words and actions is understood by the reader (and/or certain other characters), but not by the character in question. The result is often a striking sense of futility and strong feeling of tension. For example, if you know that a character has lied to her husband, and her husband takes her in his arms and tells her that the main reason he loves her is because she is such an honest person, that's irony. Or let's say the wife has just secretly packed her bags because she plans to sneak off and leave her husband in the middle of the night. When the husband in the next scene tells her how much he loves her and what a wonderful life they'll have together, that's irony. We know something he doesn't know, and that creates a certain tension within us.

As you can see from the Poe passage quoted, the **language** and **style** used by the writer also help establish the tone of a story. First, let us consider **diction**—the author's choice of words. Poe could have used the word "quiet" instead of "soundless," or "bland" instead of "dull," or "monotonous" instead of "dreary"—but he did not. Instead, he chose heavy, negative words, dreary words, words that have an emotional impact on the reader appropriate to the meaning of the story. He also chose to write a long, frequently punctuated, and very descriptive sentence.

Some authors, on the other hand, will write in short sentences and offer minimal description. Compare Poe's opening sentence to the following opening paragraph from Hernando Tellez's short story, "Just Lather, That's All":

> He said nothing when he entered. I was passing the best of my razors back and forth on a strop. When I recognized him I started to tremble. But he didn't notice. Hoping to conceal my emotion, I continued sharpening the razor. I tested it on the meat of my thumb, and then held it up to the light. At that moment he took off the bullet-studded belt that his gun holster dangled from. He hung it up on a wall hook and placed his military cap over it. Then he turned to me, loosening the knot of his tie, and said, "It's hot as hell. Give me a shave." He sat in the chair.

What do you notice about Tellez's language and style here? How does it differ from Poe's? Write your answer in the lines provided.

1. _____

2. _____

3. _____

Did you notice the brevity of most of Tellez's sentences? Can you guess why Tellez would choose to write this way? Imagine if you were the narrator. You are nervous—in fact, you start to tremble. How would you breathe? In long, deep breaths? No. Your breaths would come short and quick—just like the sentences. The words are also simple, common words, like the words of a local barber. Notice that the only object that gets detailed description is the gun holster—the weapon, presumably the object that makes the barber so nervous.

In considering an author's language and style, remember that authors choose their words carefully because each word has a **denotation** (exact, or dictionary, meaning), a **connotation** (implied or suggested meaning), and an **emotional register**. Many words also have several meanings or possible meanings, so the author's diction must be looked at carefully.

Another aspect of language and style you should be familiar with is **figurative language**. Figurative language is the use of **similes, metaphors,** and **personification**. Essentially, all figurative language is comparative. A simile, for example, *compares* Thing A to Thing B using the words *like* or *as* (Thing A is *like* Thing B). A metaphor, on the other hand, makes a stronger comparison by saying A is not *like* B; rather, A *is* B. For example: "Your eyes are *like* the deep blue sea" and "Your eyes are *as* blue *as* the sea" are similes; "Your eyes *are* the deep blue sea" is a metaphor.

Personification takes non-human or non-animal objects and endows them with human or animal characteristics. For example, if you tell someone to meet you at *the foot* of the mountain, you are using personification. Mountains don't *literally* have feet; you're speaking *figuratively* here, comparing the mountain to a person. Figurative language is a powerful writing tool that enables writers to create powerful impressions and express meaning in ways that simple description cannot.

Look at the following sentence from Toni Morrison's novel *The Bluest Eye* as an example:

> Nuns go by as quiet as lust, and drunken men and sober eyes sing in the lobby of the Greek hotel.

Did you notice the simile, "as quiet as lust"? And the personification—eyes singing? Not only does Morrison employ effective figurative language here—she has also chosen words that incorporate the action of the novel, in which Pecola's drunken father rapes her and in which Pecola dreams of having blue eyes.

Imagery is another way writers use language to create a certain effect. Imagery is the representation of sensory experience through language. Look at the following paragraph from Langston Hughes' short story "On the Road":

> The big black man turned away. And even yet he didn't see the snow, walking right into it. Maybe he sensed it, cold, wet, sticking to his jaws, wet on his black hands, sopping in his shoes. He stopped and stood on the sidewalk hunched over—hungry, sleepy, cold—looking up and down.

While you read the passage, could you almost *imagine* how the snow felt? The cold, wet, stickiness of it? Feel it on your face and in your shoes? If you could, that's because Hughes has used imagery here—sensory description—to help you *see* and *feel* just how his character felt.

The **point of view** from which a story is told is also very important in establishing tone and creating meaning. Stories are always **narrated** by someone (this "someone" is a fictional character, not the author). The narrator can be **first person** (the narrator uses the pronoun "I" and tells his or her own story), **second person** (using the pronoun "you," as if it is your, the reader's, story), and **third person** (using the pronoun "he" or "she" to tell someone else's story).

A first person narrator allows us to get directly inside his or her head; we share his/her thoughts and see the world of the story through his or her eyes. It is a very **subjective** view. A second person narrator makes the story ours—we are in it, we are acting it out. A third person narrator tells a more **objective** story because he or she describes things that happened to others, not to himself. Third person narrators are often **omniscient**, meaning they know everything about all of the characters; they can reveal to us the character's feelings and ideas.

Fiction writers often employ **symbolism** in their stories. A **symbol** is a person, place, or object invested with special meaning or significance. A flag is a good example. Practically speaking, a flag is a decorated piece of cloth. But symbolically, a flag represents a group, like a state or nation. Colors are also symbolic: white is often used to represent purity or innocence, blue to represent sadness, purple to represent royalty. Rocks are often symbols of stability, birds of freedom, trees of growth. The list is literally endless, and different authors may use the same object to represent different things. The meaning of a symbol must be determined within the story's context. In fiction, a symbol will often appear repeatedly so that it will not be overlooked.

All of these elements add up to express the story's **theme**—its overall meaning. The theme is the end result of a work: the point it has made, the questions it has raised, the position it has taken on an issue. The theme of a story is rarely explicit, so we must look at all the other elements to see what point the author is trying to make.

MAJOR MOVEMENTS IN FICTION

Storytelling originated with the development of human language. In fact, all of our first stories were not written but told, passed on from generation to generation not by books, but by word of mouth. Among the first stories were **myths,** which later developed into **parables** and **fables**. These stories all have clear morals or lessons to be learned and/or attempt to make some kind of sense of human behavior and the world around us. Stories still do that today, but forms of storytelling have evolved dramatically. Today, it is relatively rare for someone to write myths, which tend to speak of and for a whole culture. Instead, authors in general have become more interested in developing individual characters and conflicts. They are often less interested in teaching an explicit lesson than in exploring an issue, asking questions and forcing readers to think rather than providing them with specific, easy answers.

Today we can classify fiction into several categories, including **realist, romanticist, impressionist, expressionist, naturalist,** and **experimental**. These terms basically express an author's view of how the world should be represented in fiction. For example, a realist strives to tell his or her story in a way that is as true to life as possible. All aspects of human nature, no matter how base, are shown, and events are told in chronological order from a realistic, controlled, utterly objective eye. An experimental story teller, on the other hand, will literally experiment with plot. He or she may begin the story at the end, for example, and tell it backwards, to reflect how we often come to understand our lives; or tell the story from several points of view at once; or eliminate the resolution at the end of the story, since life often doesn't provide us with such a neat conclusion.

The category a writer fits into often depends on the writer's time, but it also depends on the type of fiction he or she is writing. While it is helpful to know what category a writer belongs to, it is more important simply to look for clues to meaning that the author has left in the text in the different elements of fiction. Though some authors may ask you to dig deeper than others, the answers are always there.

Myths, Parables, and Fables

Myths are stories that attempt to explain a custom, practice, belief, or natural phenomenon. For example, most societies have creation myths, stories that offer an explanation of how we came to people the earth. Many cultures also have myths to explain thunder, lightning, and other natural phenomena. Myths, in short, are ways of making sense of the world around us. The following Native American myth, "Salt Woman Is Refused Food,"* is a good example. Read it carefully, and then answer the questions below.

Old Salt Woman had a grandson, and they were very poor. They came to Cochiti and went from house to house, but people turned them away. They were all busy cooking for a feast. At that time they used no salt.

When Salt Woman and her grandson had been to all the houses, they came to a place outside the pueblo where lots of children were playing. All the children came to see the magic crystal Salt Woman had in her hand. She led them to a pinon tree and told them each to take hold of a branch of the tree and swing themselves. Using her magic crystal, she turned them into the chaparral jays who live in the pinon trees. "When we were in the pueblo, nobody would invite us to stay," Salt Woman said. "From now on you will be chaparral jays."

Salt Woman and her grandson went south and came to Santa Domingo, where they were well treated and fed. After they had eaten and were leaving, Salt Woman said, "I am very thankful for being given food to eat," and she left them some of her flesh. The people of the house ate it with their bread and meat. It tasted good—salty.

"At Cochiti," Salt Woman told them, "they treated me badly, and when I left, I took all the children outside the pueblo and changed them into chaparral jays roosting in the pinon tree. But to you I am grateful. Therefore remember that if I am in your food, it will always taste better. I will go southeast and stay there, and if any of you want more of my flesh you will find it at that place. And when you come to gather, let there be no laughing, no singing, nothing of that kind. Be quiet and clean." So she left Santo Domingo and went to Salt Lake, where we get salt today.

Questions

1. Why does Salt Woman turn the children into chaparral jays?

2. What does she leave behind for the people of Santo Domingo?

3. What natural phenomenon does this myth explain?

Answers

You can see that the people of Santo Domingo were kind to Salt Woman, and so she rewarded them with salt, which was very important for cooking and the preservation of food for Native Americans. She literally left behind a piece of her flesh (answer number 2), which they used as salt. The people of Cochiti, however, were not so kind to Old Salt Woman. In fact, they turned her away from their houses, and that's why (answer number 1) she turned their children in to chaparral jays.

This myth explains how this tribe came to use salt in a story that reflects the generous nature of the Santo Domingo people. It also shows why the Santo Domingo Indians must gather salt ceremoniously and solemnly—laughter and singing would upset the mythical Salt Woman. Finally, this myth also explains the natural phenomenon of the salt lake (answer number 3).

Parables are short **allegorical** tales that illustrate a moral or religious truth. An **allegory** is a story in which the characters symbolize ideas or values, such as faith and doubt. Their purpose is primarily **didactic**: to teach the listener a moral or religious lesson. Read the following parable from the Bible and see if you can determine the lesson:

PARABLE OF THE SOWER

A large crowd was gathering, with people resorting to him from one town after another. He spoke to them in a parable:

"A farmer went out to sow some seed. In the sowing, some fell on the footpath where it was walked on and the birds of the air ate it up. Some fell on rocky ground, sprouted up, then withered through lack of moisture. Some fell among briers, and the thorns growing up with it stifled it. But some fell on good soil, grew up, and yielded grain a hundredfold."

As he said this he exclaimed: "Let everyone who has ears attend to what he has heard." His disciples began asking him what the meaning of this parable might be.

Before we continue this parable to hear how Jesus explains it, what do you think it means? Remember that the sower and the seeds are symbolic. What might they represent? Keep in mind that Jesus is attempting to teach his followers a religious lesson.

The meaning of the parable might be:

_____

_____

_____

Now here's the rest of the parable:

> "This is the meaning of the parable. The seed is the word of God. Those on the footpath are people who hear, but the devil comes and takes the word out of their hearts lest they believe and be saved. Those on the rocky ground are the ones who, when they hear the word, receive it with joy. They have no root; they believe for a while, but fall away in time of temptation. The seed fallen among briers are those who hear, but their progress is stifled by the cares and riches and pleasures of life and they do not mature. The seed on good ground are those who hear the word in a spirit of openness, retain it, and bear fruit through perseverance." *

The New American Bible, Luke Ch. 8

Fables, much like parables, are short stories designed to teach a moral lesson or reveal a truth about human nature. Usually the characters are animals or inanimate objects instead of people. The Greek writer, Aesop, is one of the most recognized fable writers. In the Aesop fable below, and in most fables, the "moral of the story" is very clear:

THE FOX AND THE GRAPES

A famished Fox saw some clusters of ripe black grapes hanging from a trellised vine. She resorted to all her tricks to get at them, but wearied herself in vain, for she could not reach them. At last she turned away, beguiling herself of her disappointment and saying: "The Grapes are sour, and not ripe as I thought."
Moral: It is easy to despise what we cannot have.

As you can see, with myths, parables, and fables, it is usually easy to find the author's theme. In novels and short stories, however, the reader must usually work a bit harder to uncover the theme.

Novels and Short Stories

Most of us, when we read literature, read novels and short stories. The difference between these two genres is primarily a physical one: novels are much longer than *short* stories. Some short stories can be ten, twenty, even thirty pages in length, but they lack the development of plot and character you will find in a novel of, say, 100 pages or more. In a short story, on the other hand, because of its brevity, each word becomes very important. There is no room for wasted words, so each word must be chosen with care. Whereas in a novel the author has dozens of pages to describe a character, in a short story, the author has only a few pages or paragraphs. Those works that are in between—fifty or sixty pages long—are considered **novellas**, very short novels or very long short stories.

There are many different types of novels and short stories, from the picaresque (like *Don Quixote*) to the romance, from horror to science fiction. We can group these types into the following general categories: **detective/thriller, adventure/romance, science fiction/fantasy,** and **social/historical**.

Detective/thriller stories, like the famous Sherlock Holmes series and Stephen King's stories, have always enjoyed immense popularity. This is due in large part to our desire to have crimes solved and evil punished. It is also due to the fact that, when reading a detective or thriller novel, we can immerse ourselves in a world of danger without ever really putting ourselves at risk. Whenever we wish, we can simply put the book down. We also, especially when reading detective stories, have the opportunity to "be" the person solving the crime: The clues are laid out for us in the story, just as they are for the detective, and if we are logical enough and observant enough, we can occasionally solve the mystery ourselves before the detective does.

Take a look at the following passage from the first chapter of Sir Arthur Conan Doyle's *The Hound of the Baskervilles*, one of the finest murder mysteries ever written. In it, Sherlock Holmes asks the narrator, Watson, to determine a stranger's character by examining the stranger's walking stick. Watson makes certain assumptions about the stranger, Dr. Mortimer, based on his stick, but he is quickly corrected by Holmes:

"Has anything escaped me?" I asked with some self-importance. "I trust that there is nothing of consequence which I have overlooked?"

"I am afraid, my dear Watson, that most of your conclusions were erroneous. When I said that you stimulated me I meant, to be frank, that in noting your fallacies I was occasionally guided towards the truth. Not that you are entirely wrong in this instance. The man is certainly a country practitioner. And he walks a good deal."

"Then I was right."

"To that extent."

"But that was all."

"No, no, my dear Watson, not all—by no means all. I would suggest, for example, that a presentation to a doctor is more likely to come from a hospital than from a hunt, and that when the initials 'C.C.' are placed before that hospital the words 'Charing Cross' very naturally suggest themselves."

"You may be right."

"The probability lies in that direction. And if we take this as a working hypothesis we have a fresh basis from which to start our construction of this unknown visitor."

"Well, the, supposing that 'C.C.H.' does stand for 'Charing Cross Hospital,' what further inferences may we draw?"

"Do none suggest themselves? You know my methods. Apply them!'

"I can only think of the obvious conclusion that the man has practised in town before going to the country."

"I think that we might venture a little farther than this. Look at it in this light. On what occasion would it be most probable that such a presentation would be made? When would his friends unite to give him a pledge of their good will? Obviously at the moment when Dr. Mortimer withdrew from the service of the hospital in order to start in practice for himself. We know there has been a presentation. We believe there has been a change from a town hospital to a country practice. Is it, then, stretching our inference too far to say that the presentation was on the occasion of the change?"

"It certainly seems probable."

"Now, you will observe that he could not have been on the staff of the hospital, since only a man well-established in a London practice could hold such a position, and such a one would not drift into the country. What was he, then? If he was in the hospital and yet not on the staff he could only have been a house-surgeon or a house-physician—little more than a senior student. And he left five years ago—the date is on the stick. So your grave, middle-aged family practitioner vanishes into thin air, my dear Watson, and there emerges a young fellow under thirty, amiable, unambitious, absent-minded, and the possessor of a favourite dog, which I should describe roughly as being larger than a terrier and smaller than a mastiff."

Pre-GED Interpreting Literature and the Arts

Questions

1. In this passage, how does Holmes come to his conclusions about the visitor? Upon what does he base his assumptions? List the "clues" he uses to piece together a portrait of his visitor.

2. In paragraph 2, erroneous means

    (1) silly.

    (2) incorrect.

    (3) logical.

    (4) positive.

Answers

1. | Clue | Conclusion |
    |---|---|
    | Stranger is a doctor, and his walking stick has initials C.C.H. | C.C.H. stands for Charing Cross Hospital. |
    | C.C.H. = Charing Cross Hospital | The doctor got the walking stick from Charing Cross Hospital. |
    | The doctor now works in the country. | The doctor got the stick from the hospital as a gift when he left the hospital to practice in the country. |
    | The doctor is now in country practice. | The doctor could not have been on the hospital staff ("such a one would not drift into the country"). |
    | The doctor could not have been on the hospital staff but he has a stick from the hospital. | The doctor must have been a house-surgeon or a house physician (therefore not much more than a senior student). |
    | The date on the stick indicates the doctor received it only five years ago. | The doctor left the hospital only five years ago as a senior student, so he is young, not middle-aged. |
    | The doctor has moved to the country. | He is unambitious. |
    | The doctor has been presented with this stick upon leaving C.C.H. | He is amiable, well-liked by his colleagues at C.C.H. |
    | The doctor left the stick at Holmes' home. | The doctor is absent-minded. |

2. "Erroneous" means (2), incorrect. This is indicated by the context of the surrounding sentences, in which Holmes says to Watson, "not that you are entirely wrong," and in which Holmes calls Watson's conclusions "fallacies."

Notice that Holmes uses essentially the same method we as readers use when we're reading actively: **making observations** and **drawing conclusions** (**inferences**) based on those observations. Avid readers of mystery novels can, by constantly going through this process, hone their own "detective"—and therefore reading—skills.

Adventure/romance stories are equally as popular as detective/thriller stories. In an **adventure** story, the characters embark on some sort of difficult, exciting, and often extremely dangerous undertaking. They often, as in Jack London's novels and short stories, and books like *Robinson Crusoe*, must struggle against forces of nature to survive. **Romance** stories also revolve around struggle—this time against forces that desire to keep lovers apart. This force may be distance, family objections, class differences, and so on. In both of these types of stories, particularly series novels such as the Harlequin Romances, readers can expect the story to end happily, with lovers riding off into the sunset or flying off to a honeymoon. One of today's most popular romance writers is Danielle Steele, whose books are consistently on the best seller list, testifying to the popularity of this **sub-genre** (category within a larger category).

Romance and adventure stories often have the reputation of being "cheap" literature. However, many adventure and romance writers go beyond the basic romance/adventure plot to incorporate discussion of social and moral issues. James Fenimore Cooper's *The Deerslayer* is such an adventure novel. In the following passage, the protagonist, Deerslayer, has a conversation with the antagonist, Hurry, about Native Americans (as you read, remember that this novel is set during pioneer times):

"Who's talking of mortals, or of human beings at all, Deerslayer? I put the matter to you on the supposition of an Injin. I dare say any man would have his feelin's when it got to be life or death, ag'in another human mortal; but there would be no such scruples in regard to an Injin; nothing but the chance of his hitting you, or the chance of your hitting him."

"I look upon the red men to be quite as human as we are ourselves, Hurry. They have their gifts, and their religion, it's true; but that makes no difference in the end, when each will be judged according to his deeds, and not according to his skin."

"That's downright missionary, and will find little favor up in this part of the country, where the Moravians don't congregate. Now, skin makes the man. This is reason; else how are people to judge of each other? The skin is put on, over all, in order that when a creatur', or a mortal, is fairly seen, you may know at once what to make of him. You know a bear from a hog, by his skin, and a gray squirrel from a black."

"True, Hurry," said the other, looking back and smiling; "nevertheless, they are both squirrels."

"Who denies it? But you'll not say that a red man and a white man are both Injins?"

"No; but I *do* say they are both men. Men of different races and colors, and having different gifts and traditions, but, in the main, with the same natur'. Both have souls, and both will be held accountable for their deeds in this life."

Questions

1. What are Hurry's ideas about Native Americans? Why?

2. What are Deerslayer's ideas about Native Americans? Why?

3. Which character's ideas do you agree with? Why?

4. Who do you think the author wants you to agree with? Why do you think so?

Answers

1. Hurry thinks that Native Americans (whom he calls "Injins") are not even human beings. We can tell this from his opening sentence, where he says, "Who's talking of mortals, or of human beings at all, Deerslayer?" He also thinks that white people like him should have no scruples (moral beliefs preventing one from doing wrong) about killing "Injins" since they are not human and they are just waiting to hurt white people. Hurry judges "Injins" as different from human beings because of the color of their skin. He thinks that once you see the color of someone's skin, you know all there is to know about that person ("Now," he says, "skin makes the man," and once you see the skin, "you may know at once what to make of him.")

2. Deerslayer sees Native Americas as being "quite as human as we are ourselves," even though they may have different customs and beliefs. Where Hurry sees difference, Deerslayer sees sameness. Deerslayer also believes that men should be judged not by their skin color but by their actions.

3. Answers will vary.

4. The author wants us to agree with Deerslayer. This is suggested primarily because Deerslayer wins this argument using Hurry's own logic. Hurry argues that "skin makes the man. This is reason; else how are people to judge of each other?" He uses animals as an example: "You know a bear from a hog, by his skin, and a gray squirrel from a black." Deerslayer turns this logic back on Hurry by pointing out, quite reasonably, that "nevertheless, they are both squirrels," which confirms Deerslayer's argument that white men and Native Americans are both men. In addition, Deerslayer speaks somewhat more eloquently than Hurry, indicating that he is more educated.

Science fiction includes stories that are based on actual or imaginary developments and discoveries in the sciences, especially technology. Many science fiction novels and short stories are set in the future—and often that future is bleak. Generally, the themes of these stories concern human use and misuse of scientific knowledge. Whether the stories are realistic or fantastic (for example, set in another universe or involving alien species), they offer a very real commentary on human nature and power. **Fantasy** stories, on the other hand, are generally set in imaginary worlds peopled with mythic creatures who have magical powers (sorcerers and the like).

One of the most famous science fiction writers is Ray Bradbury, author of numerous short stories and the classic novel *Fahrenheit 451*. The novel, published in 1953, tells of a time when books are absolutely forbidden, and any house where books are found is burned to the ground. In the following passage, the protagonist, Montag, is forced to burn his own house down after he has been caught reading and hiding books:

"Was it my wife turned in the alarm?"

Beatty nodded. "But her friends turned in an alarm earlier that I let ride.

One way or the other, you'd have got it. It was pretty silly, quoting poetry around free and easy like that. It was the act of a silly damn snob. Give a man a few lines of verse and he thinks he's the Lord of all Creation. You think you can walk on water with your books. Well, the world can get by just fine without them. Look where they got you, in slime up to your lip. If I stir the slime with my little finger, you'll drown!"

Montag could not move. A great earthquake had come with fire and leveled the house and Mildred was under there somewhere and his entire life was under there and he could not move. The earthquake was still shaking and falling and shivering inside him and he stood there, his knees half bent under the great load of tiredness and bewilderment and outrage, letting Beatty hit him without raising a hand.

"Montag, you idiot, Montag, you damn fool; why did you *really* do it?"

Montag did not hear, he was far away, he was running with his mind, he was gone, leaving this dead soot-covered body to sway in front of another raving fool.

"Montag, get out of there!" said Faber.

Montag listened.

Beatty struck him a blow on the head that sent him reeling back. The green bullet in which Faber's voice whispered and cried fell to the sidewalk. Beatty snatched it up, grinning. He held it half in, half out of his ear.

Montag heard the distant voice calling, "Montag, you all right?"

Beatty switched the green bullet off and thrust it in his pocket. "Well—so there's more here than I thought. I saw you tilt your head, listening. First I thought you had a Seashell. But when you turned clever later, I wondered. We'll trace this and drop it on your friend."

"No!" said Montag.

He twitched the safety catch on the flame thrower.

[…]

Beatty grinned his most charming grin. "Well, that's one way to get an audience. Hold a gun on a man and force him to listen to your speech. Speech away. What'll it be this time? Why don't you belch Shakespeare at me, you fumbling snob? 'There is no terror, Cassius, in your threats, for I am arm'd so strong in honesty that they pass me as an idle wind, which I respect not!' How's that? Go ahead now, you second-hand litterateur, pull the trigger." He took one step toward Montag.

Montag only said, "We never burned *right* ..."

"Hand it over, Guy," said Beatty with a fixed smile.

And then he was a shrieking blaze, a jumping, sprawling, gibbering mannikin, no longer human or known, all writhing flame on the lawn as Montag shot one continuous pulse of liquid fire on him. There was a hiss like a great mouthful of spittle banging a red-hot stove, a bubbling and frothing as if salt had been poured over a monstrous black snail to cause a terrible liquefaction and a boiling over of yellow foam. Montag shut his eyes, shouted, shouted, and fought to get his hands at his ears to clamp and cut away this sound. Beatty flopped over and over and over, and at last twisted in on himself like a charred wax doll and lay silent.

[…]

The blowing of a single autumn leaf.

He turned and the Mechanical Hound was there.

It was half across the lawn, coming from the shadows, moving with such drifting ease that it was like a single solid cloud of black-gray smoke blown at him in silence.

It made a single last leap into the air coming down at Montag from a good three feet over his head, its spidered legs reaching, the procaine needle snapping out its single angry tooth.

Questions

1. What is the "green bullet" in Montag's ear?

2. How does Montag kill Beatty?

3. What is the Mechanical Hound trying to do to Montag?

4. What elements identify this as a work of science fiction?

5. Did you notice the figurative language? Can you find several similes and metaphors?

Answers

1. The "green bullet" is a communication device that had been inserted in Montag's ear, allowing him to communicate secretly with someone named Faber. Faber tells Montag through the bullet, "Montag, get out of there!" And when Montag tilts his head to listen to Faber, Beatty notices and suspects that Montag may have such a device hidden in his ear.

2. Montag kills Beatty by torching him with liquid fire from the flame thrower.

3. The Mechanical Hound is trying to inject Montag with procaine, an anesthetic.

4. Several elements identify this as science fiction: first, the futuristic scenario in which firemen do not put out fires, but start them, and in which books are illegal; second, the green bullet; and third, the Mechanical Hound.

5. There are several similes and metaphors. The avalanche and earthquake are metaphors for the upheaval in Montag's mind now that he has been caught; while Beatty is burning there is a "hiss like a great mouthful of spittle banging a red-hot stove" (simile) and a "bubbling and frothing as if salt had been poured over a monstrous black snail" (simile); Beatty "twisted in on himself like a charred wax doll" (simile); and the Mechanical Hound "was like a single solid cloud of black-gray smoke" (simile).

Finally, the fourth general category of stories is the **social/historical**. This, of course, is a very broad—and extremely important—category. Stories that deal with historical events, like war and slavery, and social issues, like prejudice and politics, all fit into this category. Of course, a good mystery, romance, or science fiction story will also deal with these issues, but they are in separate categories largely because of the plot that drives them forward (e.g., a murder that must be solved or lovers that must unite).

One group of social/historical stories of great importance today is war stories such as Erich Maria Remarque's *All Quiet on the Western Front*, about World War I, and Kurt Vonnegut's *Slaughterhouse Five*, about World War II. Many novels and stories have also been written about the American Revolution, the American Civil War (Stephen Crane's *Red Badge of Courage*, for example), the French Revolution, and the Vietnam War. Based on actual historical events, but peopled with fictitious characters, these stories raise important issues about the nature of war and humanity. Read, for example, the following passage from *All Quiet on the Western Front*:

I am often on guard over the Russians. In the darkness one sees their forms move like sick storks, like great birds. They come close up to the wire fence and lean their faces against it; their fingers hook round the mesh. Often many stand side by side, and breathe the wind that comes down from the moors and the forest.

They rarely speak and then only a few words. They are more human and more brotherly towards one another, it seems to me, than we are. But perhaps that is merely because they feel themselves to be more unfortunate than us. Anyway the war is over so far as they are concerned. But to wait for dysentery is not much of a life either.

The Territorials who are in charge of them say that they were much more lively at first. They used to have intrigues among themselves, as always happens, and it would often come to blows and knives. But now they are quite apathetic and listless. [...]

They stand at the wire fence; sometimes one goes away and then another at once takes his place in the line. Most of them are silent; occasionally one begs a cigarette butt.

I see their dark forms, their beards move in the wind. I know nothing of them except that they are prisoners; and that is exactly what troubles me. Their life is obscure and guiltless;—if I could know more of them, what their names are, how they live, what they are waiting for, what are their burdens, then my emotion would have an object and might become sympathy. But as it is I perceive behind them only the suffering of the creature, the awful melancholy of life and the pitilessness of men.

A word of command has made these silent figures our enemies; a word of command might transform them into our friends. At some table a document is signed by some persons whom none of us knows, and then for years together that very crime on which formerly the world's condemnation and severest penalty fall, becomes our highest aim. But who can draw such a distinction when he looks at these quiet men with their childlike faces and apostles' beards. Any non-commissioned officer is more of an enemy to a recruit, any schoolmaster to a pupil, than they are to us. And yet we would shoot at them again and they at us if they were free. (169–71).

Questions

1. What is the narrator's relationship to the Russians?

2. How does the narrator feel about the Russians? How can you tell?

3. What do you think is the narrator's attitude toward war? How can you tell?

4. What does <u>listless</u> mean in the third paragraph?
   (1) Lazy
   (2) Energetic
   (3) Crazy
   (4) Without energy
   (5) Dangerous

Answers

1. The narrator is an enemy of the Russians. He is sometimes a guard over them in a prisoner of war camp.

2. We can infer that the narrator does not feel hatred for the "enemy." Instead, he feels some compassion. Even though he claims that

he does not know enough about them to feel "sympathy," he does feel sorry for them and troubled by the fact that they are prisoners. As far as he knows, the Russians are "guiltless." Because he does not know the details of their lives, his pity is general: He sees in them "the suffering of the creature, the awful melancholy of life and the pitilessness of men." He considers them "more human and more brotherly towards one another" than he and his men are. We can also tell he has compassion for these men because of his diction (choice of words). He describes them as "quiet" men with "child-like" (therefore innocent) faces and "apostles' beards," suggesting that not only are these men innocent of any crimes, but they are also holy in their suffering and undeserving of such treatment.

3. We can infer that the narrator is disgusted by war, which he sees as a game whose rules change at the whims of a few remote leaders. This is evident in the last paragraph, where the narrator says, "A word of command has made these silent figures our enemies; a word of command might transform them into our friends." The nameless leaders "whom none of us knows" sign documents that change friends to enemies and enemies to friends. These leaders and commanders are "more of an enemy" to the soldiers, the narrator says, than these Russians are to the narrator. And yet, because of a command, they would try to kill each other if the Russians were not imprisoned.

4. **(4)** We can tell from the context that "listless" means (4), without energy. The Territorials said the Russians "were much more lively at first"—they had energy and even had "intrigues" among themselves that led to violent fights. Now, however, they have become apathetic (caring about nothing) and listless (without vigor, dull, indifferent).

Now take a look at the following excerpt from John Steinbeck's *In Dubious Battle*, a social novel set in America in the 1920s. A group of farm workers is striking. See if you can determine why they are striking and what Doc Burton is doing there.

"Then you don't think the cause is good?"

Burton sighed. "You see? We're going to pile up on that old rock again. That's why I don't like to talk very often. Listen to me, Mac. My senses aren't above reproach, but they're all I have. I want to see the whole picture—as nearly as I can. I don't want to put on the blinders of 'good' and 'bad,' and limit my vision. If I used the term 'good' on a thing I'd lose my license to inspect it, because there might be bad in it. Don't you see? I want to be able to look at the whole thing."

Mac broke in heatedly, "How about social injustice? The profit system? You have to say they're bad."

Dr. Burton threw back his head and looked at the sky. "Mac," he said. "Look at the physiological injustice, the injustice of tetanus, the injustice of syphilis, the gangster methods of amoebic dysentery—that's my field."

"Revolution and communism will cure social injustice."

"Yes, and disinfection and prophylaxis will prevent the others."

"It's different, though; men are doing one, and germs are doing the other."

"I can't see much difference, Mac."

"Well, damn it, Doc. there's lockjaw every place. You can find syphilis in Park Avenue. Why do you hang around with us if you aren't for us?"

"I want to *see*," Burton said. "When you cut your finger, and streptococci get in the wound, there's a swelling and a soreness. That swelling is the fight your body puts up, the pain is the battle. You can't tell which one is going to win, but the wound is the first battleground. If the cells lose the first fight the streptococci invade, and the fight goes on up the arm. Mac, these little strikes are like the infection. Something has got into the men; a little fever has started and the lymphatic glands are shooting in reinforcements. I want to see, so I go to the seat of the wound."

"You figure the strike is a wound?"

"Yes. Group-men are always getting some kind of infection. This seems to be a bad one. I want to *see*, Mac. I want to watch these group-men, for they seem to me to be a new individual, not at all like single men. A man in a group isn't himself at all, he's a cell in an organism that isn't like him any more than the cells in your body are like you. I want to watch the group, and see what it's like. People have said, 'mobs are crazy, you can't tell what they'll do.' Why don't people look at mobs not as men, but as mobs? A mob nearly always seems to act reasonably, for a mob."

"Well, what's this got to do with the cause?"

"It might be like this, Mac: When group-man wants to move, he makes a standard. 'God wills that we re-capture the Holy Land'; or he says, 'We fight to make the world safe for democracy'; or he says, 'We will wipe out social injustice with communism.' But the group doesn't care about the Holy Land, or Democracy, or Communism. Maybe the group simply wants to move, to fight, and uses these words simply to reassure the brains of individual men. I say it *might* be like that, Mac."

"Not with the cause, it isn't," Mac cried.

(147–8)

Questions

1. What cause is Mac fighting for? What is he hoping to "cure"?

2. Notice how Dr. Burton compares the striking men to a wound. To what does he compare group-men?

3. What does Dr. Burton think is the difference between group-men and individual men?

4. Why is Dr. Burton there with the striking men? Do you think he believes in the cause?

Answers

1. Mac is fighting to cure "social injustice." The "cause" is communism, specifically a communist revolution that he hopes will cure social injustice.

2. Dr. Burton compares group-men to a new type of living organism. Each individual man in the group is a cell in that organism.

3. Group-men are very different organisms from an individual man. Group-men, according to Dr. Burton, "are always getting some kind of infection," but the "cure" group-men seek is merely an excuse for action. Group-men, Dr. Burton says, don't care about the "cause"—the group simply "want[s] to move, to fight," and this cause is what "reassure[s] the brains of individual men" that they're doing the right thing.

4. Dr. Burton is there because he wants to analyze this social and political "illness" (the strike) like he would examine a physical illness. He wants to see just what is causing the "infection," so he has gone to "the seat of the wound." He is not there as a believer in the cause, but as an observer. It seems that he does not believe in the cause; rather, he wants to test the validity of his theory about group-men.

NON-FICTION

Unlike fictional prose, **non-fictional** texts are not peopled by imaginary characters who act out imaginary events. Instead, the subject of non-fictional prose is "reality"—real people and real events. In non-fiction texts, the author's task is to describe and comment on those people and events.

ELEMENTS OF NON-FICTION

Because non-fiction is generally more straightforward than fiction, there are fewer elements to consider—but we still must consider them carefully. The basic elements of non-fiction are:

- voice;
- language and style;
- structure;
- tone;
- theme.

Unlike fiction, where the story is always told through an intermediary (a narrator of some sort), in non-fiction the person telling the "story" is the author him/herself. There is no "filter" here; the author speaks his or her mind directly to the reader. Thus, the **voice** is that of the author. But there are as many different voices as there are authors.

Non-fiction authors combine their voice with a **point of view** that will best express their ideas. This point of view helps establish the appropriate relationship to the reader. For example, an essay that uses "I" creates a certain intimacy between the author and reader, for the writer is sharing ideas and feelings directly with the reader. But there are many different levels of intimacy. The "I" can be the voice of a colleague at a staff meeting or it can be the voice of an intimate friend. When an author uses "you," on the other hand, he or she wants you to feel as if you are taking part in the action or ideas he or she is describing. The use of "he" or "she," in contrast, creates a certain distance and implies objectivity.

The **language and style** of an essay helps control the **tone** of the text. For example, an author may use formal or informal language and write in long or short sentences. The long, formal sentences, of course, have a different voice than short, informal sentences. The author may use **jargon** (technical, specialized language) or slang as well. His or her diction can help create a tone of voice that whispers in our ears or shouts from a pulpit. Compare, for example, the language and style and the resulting tone of voice in these two essay excerpts. Read the passages out loud so you can literally hear how the voices sound.

From E.B. White's "The Essayist" (1977):

> The essayist is a self-liberated man, sustained by the childish belief that everything he thinks about, everything that happens to him, is of general interest. He is a fellow who thoroughly enjoys his work, just as people who take bird walks enjoy theirs. Each new excursion of the essayist, each new "attempt," differs from his last and takes him into new country. This delights him. Only a person who is congenitally self-centered has the effrontery and the stamina to write essays.

From Joy Williams' "Save the Whales, Screw the Shrimp" (1988):

> *Acid rain!* And we all know what this is. Or most of us do. People of power in government and industry still don't seem to know what it is. Whatever it is, they say, they don't want to curb it, but they're willing to study it some more. Economists call air and water pollution "externalities" anyway. Oh, acid rain. You do get so sick of hearing about it. The words have already become a white-noise kind of thing.

Questions

1. Which writer uses more a formal language and style?

2. How would you describe White's voice? Is it intimate or distanced? Professional or casual? Haughty or humble? Critical or complimentary?

3. How would you describe Williams' voice?

Answers

1. White uses more formal language and style. We can tell from the vocabulary, from the sentence structure, and from the point of view. (See 2 and 3, below.)

2. White's voice is distant. Although he is talking about himself (the essayist), he uses the third person point of view ("he"). He is professional and formal, using words like "excursion," "congenitally," and "effrontery" instead of simpler, more accessible words. He is obviously critical of the essayist who is "childish" and "self-centered," but his criticism is somewhat tongue-in-cheek since he himself is an essayist. The mockery of the essayist is light because it is balanced by positive terms like "self-liberated" and "stamina."

3. Williams uses the first-person point of view ("we," "us") and second-person point of view ("you") to close the distance that White purposely creates in his essay. Williams is very expressive, using italics and an exclamation point in the very first sentence. Her style is very casual, like a conversation—it is fragmentary ("Oh, acid rain") and includes slangy phrases like "a white noise kind of thing." Rather than trying to sound like a critic, Williams is trying to sound like a friend.

The **structure** of a non-fictional text—particularly an essay—is also important. Is it divided into sections, or chapters? Why or how? Upon what principle are these divisions made? What's the logic? "The Knife," for example, is divided into three sections. The first section, which describes the knife, is controlled by three different pronouns: "one," "it," and "I." "It" is the knife, and "I" is Selzer, an experienced surgeon. Section two is about the patient under the knife in surgery, and Selzer has chosen to call the patient "you":

> "And what of that *other*, the patient, you, who are brought to the operation room on a stretcher, having been washed and purged and dressed in a white gown? Fluid drips from a bottle into your arm, diluting you, leaching your body of its personal brine. As you wait in the corridor, you hear, from behind the closed door the angry clang of steel upon steel, as though a battle were being waged."

In the third section of "The Knife," when the operation takes place, the surgeon is no longer "I" but "he." In this section, the third-person pronoun dominates to reflect the objectivity and distance there must be between surgeon and patient during the operation (there is no room for intimacy when the patient is on the operating table).

Many non-fiction writers, especially essayists and journalists, **frame** their texts—that is, they start with an idea or incident in the beginning and return to that idea or incident in the end. For example, look at the first and last paragraphs of another Selzer essay, "The Exact Location of the Soul":

The first sentence of the first paragraph:

> "Someone asked me why a surgeon would write. Why, when the shelves are already too full?"

The first sentence of the last paragraph:

> "Did you ask me why a surgeon writes? I think it is because I wish to be a doctor."

By framing a text, an author physically brings us full circle by bringing us back to the idea that he or she started with. It helps the reader recall all that came before in the essay and provides a very strong sense of closure.

Carefully examine the structure of the non-fictional texts you read. The way an author divides and organizes his or her essay is always deliberate. Each division is made for a reason; perhaps there is a shift in subject, tone, or voice, as in Selzer's "The Knife." The order in which an author presents his or her ideas is also significant, because authors usually (but, of course, not always) save their most important ideas and most powerful examples for last.

THE ESSAY

You probably had to write many essays throughout your years at school, and you may even have thought it a dreadful task. But the essay is a truly wonderful form of non-fiction prose that is informative, interesting, and often innovative.

The word *essay* comes from the Latin word *exigere*, meaning "to weigh, to examine." And that's what essays do—examine an author's opinion. Most essays are **expository**. They make an **assertion** (a statement or declaration that requires evidence or explanation to be accepted as true) about their subject and then support that assertion with concrete "evidence." The type of evidence an author provides depends largely on the type of essay he or she is writing. A **personal essay**, for example, will support its **assertion (thesis)** with evidence from the author's personal experience. **Exposition** is essentially the act of setting forth or explaining facts or ideas in detail. Thus, the expository essay.

The key to understanding essays is determining the author's **thesis**—the main assertion that governs the whole essay. Because the thesis encompasses the whole essay, it is usually (but not always) located at the beginning of the essay. A thesis is a statement that describes both the *subject* of the essay and the author's *position* on that subject. It is not, therefore, simply an announcement of the subject ("This essay is about the causes of World War II"). It is also not a command ("Let's examine the causes of World War II"), nor is it a question ("What were the causes of World War II?") or a fact ("There were many causes of World War II"). A thesis must put forth an opinion, take a position—express the author's attitude. The following sentence, then, is a thesis statement: "We must understand the causes of World War II to prevent a World War III."

A general formula to remember is the following: **thesis = subject + attitude.** Now look at the sentences below. Which of the following is a thesis statement? Which sentence makes an assertion?

1. This summer I went to Tasmania.
2. Tasmania is an island south of Australia.
3. What is there for a tourist to do in Tasmania?
4. Tasmania may be the most fascinating place in the world.
5. In this paper I will describe my trip to Tasmania.

The correct answer, of course, is number four. It is the only sentence that expresses the author's attitude. Sentences one and two are facts; sentence three is a question; and sentence five is an announcement.

For this essay about Tasmania to be successful, the author must explain *how* or *why* Tasmania is so fascinating by providing "evidence" for this assertion. He or she might, for example, describe the people of Tasmania, their customs, the history, the social norms, the tourist attractions, the climate—whatever makes Tasmania a fascinating place to the author.

These are the **supporting ideas** of the essay, and the more detailed they are, the more convinced the readers will be that Tasmania is, indeed, a fascinating place.

Essentially, then, the purpose of any essay is to convince—to convince the readers of the validity of the thesis.

Look at the following excerpt from "Walden" by Henry David Thoreau. See if you can identify Thoureau's thesis and the ideas that support it. Don't forget to mark up the text and make observations as you go.

> Why should we be in such desperate haste to succeed, and in such desperate enterprises? If a man does not keep pace with his companions, perhaps it is because he hears a different drummer. Let him step to the music which he hears, however measured or far away. It is not important that he should mature as soon as an apple-tree or an oak. Shall he turn his spring into summer? If the condition of things which we were made for is not yet, what were any reality which we can substitute? We will not be shipwrecked on a vain reality. Shall we with pains erect a heaven of blue glass over ourselves, though when it is done we shall be sure to gaze still at the true ethereal heaven far above, as if the former were not?
>
> However mean your life is, meet it and live it; do not shun it and call it hard names. It is not so bad as you are. It looks poorest when you are richest. The fault-finder will find faults even in paradise. Love your life, poor as it is. You may perhaps have some pleasant, thrilling, glorious hours, even in a poorhouse. The setting sun is reflected from the windows of the alms-house as brightly as from the rich man's abode; the snow melts before its door as early in the spring. I do not see but a quiet mind may live as contentedly there, and have as cheering thoughts, as in a palace.
>
> Rather than love, than money, than fame, give me truth. I sat at a table where were rich food and wine in abundance and obsequious attendance, but sincerity and truth were not, and I went away hungry from the inhospitable board. The hospitality was as cold as the ices. I thought that there was no need of ice to freeze them. They talked to me of the age of the wine and the fame of the vintage; but I thought of an older, newer, and purer wine, of a more glorious vintage, which they had not got, and could not buy. The style, the house and the grounds and "entertainment" pass for nothing with me. I called on the king, but he made me wait in his hall, and conducted like a man incapacitated for hospitality. There was a man in my neighborhood who lived in a hollow tree. His manners were truly regal. I should have done better had I called on him.

Questions

1. What do you feel is the main idea of the passage? Can you find statements that support it?

2. Did you notice what imagery the author uses the most?

3. What do you think the author is speaking out against in the first paragraph?

4. What is the main point of the third paragraph?

Answers

1. The main idea of the passage is that people should not get caught up in the pursuit of material wealth since it does not make them truly "rich" and that they should learn to love

their lives and appreciate the things that they have. Statements such as "It is not important that we mature like an apple-tree or an oak," or "However mean your life is, meet it and live it," support the main idea.

2. The author uses mostly images of nature. "Apple-tree or an oak," "the snow melts," or "setting sun" are just some of the images of nature that the author uses.

3. "Desperate haste to succeed" as well as "keeping pace with companions" are two of a number of phrases that suggest society pressures man into a quest for achievement and success. Thoreau is speaking out against giving into society's pressures of making men do what society wants, not what a man really wants to do ("let him step to the music he hears, perhaps it is because he hears a different drummer").

4. The third paragraph stresses that truth is what makes us rich, not the material things we have.

Sometimes, of course, the author's thesis may not be explicit and we may have to draw our own conclusions based on the examples and ideas expressed in the essay. If there is no clear thesis statement, there will be clues to help you. The main idea of an essay, for example, is often suggested by its title. The words the author uses to describe his or her subject also provide clues. Are they complementary or negative? Light or heavy? You can also look to the conclusion of the essay and ask yourself: "What does this essay add up to? What statement can I make that will sum up all of the ideas and examples expressed here?" If you read actively and look for clues, you should be able to find the author's thesis.

Now read the following editorial essay "What Army Ads Don't Say" by Jessica Siegel. Unlike Thoreau, Siegel does not provide a clear thesis statement. However, Siegel's main idea should be clear to you if you add up her clues.

WHAT ARMY ADS DON'T SAY

When Baghdad was bombed, I thought of the kids I taught for 10 years in Seward Park High School on Manhattan's Lower East Side. They're the sons and daughters of the working poor and the welfare system. Given their stunted job expectations—McDonalds, supermarkets, the drug trade—no wonder the military is alluring.

"Be All You Can Be," ads say on television during football and basketball games, on the radio, on billboards. "It's Not a Job, It's an Adventure." I was enraged watching kids buy the targeted siren song of discipline, travel, skills and, of course, the inevitable "training." Training for what, really, in the end? The recruiters didn't get around to surgical strikes and limited incursions.

When a basketball player from Avenue D told me in loving detail about the drawer of underwear, perfectly folded in triangles, in the room of an Army friend he visited, what was he thinking about? His story told as much about what he wanted to escape from as what he wanted to escape to.

Military recruiters come to high schools' college fairs, muscles bulging, in sharp uniforms with knife-sharp creases in their trousers and mirror-like shoes. (College recruiters don't look that good.) They hand out posters, book covers, promises. "Just put your name down, and I'll send you more information." The mail soon brings free wallets, free socks, more glossy literature. The students get calls from recruiters, who visit them at home and play up to their families. Just sign on the bottom line....

Listening to TV coverage of the bombing, I heard that no chief executive officers of the top 20 corporations have children in the desert. This is not surprising. Army, Navy, Marine and Air Force ads in high school newspapers in the inner city fetch more attention than they do in papers in private schools and affluent suburbs. Remembering Vietnam, I wasn't surprised when I heard that a third of our troops in Saudi Arabia awaiting an Iraqi counterattack are African Americans and Hispanics.

The Army has always provided an escape for the poor—a place where they got three meals and a bed. But today, given the horrors of the inner city, racism and limited opportunities, it is especially attractive.

So often at Seward Park, I read sensitively written essays. When I asked the writer if he was going to college, he'd grow quiet and say: "I think I'll join the military. They say they'll give you training." No recruiter, no ad, ever used the word "casualties."

Questions

1. How does the military entice recruits?

2. Why is the military so attractive to inner city kids?

3. What do you think is Siegel's main idea? Write a thesis statement that expresses this idea. Be sure your statement is general enough to encompass the entire essay.

4. How does Siegel support this thesis?

5. What does <u>affluent</u> (paragraph 5) mean?

   (1) Strange

   (2) Wealthy

   (3) Private

   (4) Distant

   (5) Educated

Answers

1. The military officers entice recruits with their sharp, impressive appearance, their slogans, and their gifts, such as posters, book covers, wallets, and socks, as well as phone calls and visits. Most importantly, perhaps, they entice recruits with promises of a better future through "training."

2. The military is so attractive to inner city kids because in their environment they have very limited job opportunities, so the prospect of traveling and "training" is very appealing. Also, inner city youths often grow up in environments that lack discipline and structure, which many of them crave. Furthermore, the military offers the certainty of three meals a day and a place to sleep, as well as the opportunity to escape the cycle of poverty.

3. A good thesis statement for this essay might be: *The military's recruiting of inner city youths is deceitful,* or, *The military recruiters talk about opportunities, but never about the opportunity to die.*

4. Siegel supports this thesis by listing the techniques that recruiters use to attract inner city youths while pointing out that nowhere in that recruiting process is the possibility of death discussed. She also supports this thesis by relating her experiences with inner city students.

5. "Affluent" suburbs are (2), wealthy suburbs. This is suggested by the context of the paragraph, which talks about the children of wealthy chief executive officers and expensive

private schools (in contrast to the inner city schools, where poor children attend free, often lower-quality public schools).

Types of Essays

When we talk about essays, we often break them down into several categories called **rhetorical modes: description, narration, illustration, process, classification, compare and contrast, persuasion,** etc. However, all essays are essentially **argument** essays. Whether the author is describing his or her hometown or arguing about abortion, he or she is making an assertion and supporting it. To say "Everyone should grow up in a hometown like mine" is the same as to say "Abortion should [or should not] be legal." In both cases the writer is making an assertion and must support it.

The rhetorical mode of an essay is determined by its purpose. If my main intent is to describe my hometown, then I am writing a **descriptive** essay. If my main intent is to tell you a story about something that I experienced, then I am writing a **narrative** essay. (Notice I am still writing to convince you of something—the beauty of my hometown, for example, or the importance of this experience in my life.) If my main intent is to persuade you that my position on an issue is the most reasonable one, I am writing a **persuasive** essay.

One form of literature, particularly in nonfiction essays, is **satire**. A satire is a work that exposes and ridicules someone or something, particularly human vices and follies, in the hope of bringing about change. A satire relies heavily on **verbal irony** (when the intended meaning is the opposite of the expressed meaning). **Hyperbole**, or extreme exaggeration, is another technique often used by satirists, as well as **wit, sarcasm, understatement, reversal,** and **parody** (imitation of a well-known work for comic effect). Satire can be gentle, but it is often quite harsh, since satirists hold their subjects up for ridicule to make their point.

Jonathan Swift's "A Modest Proposal," written in 1729, is one of the most famous satirical essays. In it, Swift proposes that the Irish begin eating their infants to prevent "the children of poor people in Ireland from being a burden to their parents or country, and for making them beneficial to the public." Of course, Swift is not serious. He most certainly does not advocate cannibalism. His essay is really a vicious attack on the British government which was oppressing the Irish, particularly the poor Catholic peasants, who often had many children.

To see Swift's satire at work, read the following sentence:

> "I have been assured by a very knowing American of my acquaintance in London, that a young healthy child well nursed is at a year old a most delicious, nourishing, and wholesome food, whether stewed, roasted, baked or boiled; and I make no doubt that it will equally serve in a fricassee or a ragout."

You can see by Swift's use of **hyperbole** and by the outrageousness of his idea that he certainly does not mean to cook children. The absurdity of his assertion attests to the absurdity, in Swift's view, of British rule in Ireland at the time.

BIOGRAPHY AND AUTOBIOGRAPHY

Biography

A **biography** is the true account of a person's life. Biographers use a variety of sources to write their biographies. Ideally, their primary (main) source of information is the person who is the subject of their biography. If that person is dead or reclusive, however, biographers must rely on other sources, such as diaries, letters, and the friends, relatives, and other acquaintances and records of their subject's life.

One thing a biographer must be careful of is biased or unreliable sources—people, for example, who for some reason might not want to portray the subject's life accurately. It is important, therefore, for a biographer to do as much research as possible to verify the truth of the information he or she receives from sources.

Of course, most of us won't read a biography about just anybody. We will, however, be interested in the lives of famous or important people: geniuses like Einstein and DaVinci, celebrities like Elizabeth Taylor and Michael Jackson, innovators like Thomas Edison and Benjamin Franklin, politicians like Abraham Lincoln and Richard Nixon. In a biography, we can expect to learn about the important people, places, decisions, and events that shaped the subject's life.

In the following excerpt from *Brecht: The Man and His Work*, a biography about the master playwright Bertolt Brecht, notice how the biographer, Martin Esslin, explains a critical event in Brecht's life:

> In 1916 he left school and moved to Munich, where he began to read medicine and science at the university. But after a few terms he had to interrupt his studies. It was wartime and he was called up. Being a medical student, he became a medical orderly in a military hospital.
>
> There can be no doubt that this was one of the decisive events of his life.
>
> I was mobilized in the war [he told Sergei Tretyakov] and placed in a hospital. I dressed wounds, applied iodine, gave enemas, performed blood-transfusions. If the doctor ordered me: "Amputate a leg, Brecht!" I would answer, "Yes, Your Excellency!" and cut off the leg. If I was told: "Make a trepanning!" I opened the man's skull and tinkered with his brains. I saw how they patched people up in order to ship them back to the front as soon as possible.
>
> Seeing human beings cut up in this way and having to do the gruesome work himself was a traumatic experience that left lasting traces in Brecht's character and work. His poetry is haunted by images of dismembered bodies....

Esslin asserts that Brecht's service in the war was "one of the decisive events in his life," and then he supports that assertion with a quote from Brecht himself (this quote came from an interview between Brecht and Tretyakov).

Autobiography

An **autobiography**, on the other hand, is the story of someone's life told by that person him or herself rather than by an outside biographer. An autobiography allows us a special insight into the writer's life since the story is told from that person's own point of view. Furthermore, we can assume that someone who writes an autobiography has a special need to tell his or her story. Perhaps the author wishes to clarify certain facts about his or her life, or to teach others lessons he or she has learned, or to make a statement about a certain period of history. Moreover, in an autobiography, we can read about events and feelings that might not be uncovered by a biographer.

An autobiography, unlike a biography, is completely subjective—we have only the writer's point of view regarding the people, places, decisions, and events that make up his or her life. As a result, the story we are being told is only one version of the truth, no matter how objective the writer might try to be. As one-sided and limited as this may be, it is nevertheless deeply revealing and fascinating. An autobiography allows us to get an in-depth look at a fascinating person's life straight from the source.

In an autobiography, look for the author's idea about what people, places, and events he or she feels has shaped his or her life. The following passage from Richard Wright's narrative autobiography, *Black Boy (American Hunger): A Record of Childhood and Youth* is an example:

> "What's the matter with you?" I asked.
> Griggs glared at me, then laughed.
> "I'm teaching you how to get out of white people's way," he said.
> I looked at the people who had come out of the store; yes, they were *white*, but I had not noticed it.
> "Do you see what I mean?" he asked. "White people want you out of their way." He pronounced the words slowly so that they would sink into my mind.
> "I know what you mean," I breathed.
> "Dick, I'm treating you like a brother," he said. "You act around white people as if you didn't know that they were white. And they *see* it."
> "Oh, Christ, I can't be a slave," I said hopelessly.
> "But you've got to eat," he said.
> "Yes, I've got to eat."
> "Then start acting like it," he hammered at me, pounding his fist in his palm. "When you're in front of white people, *think* before you act, *think* before you speak. Your way of doing things is all right among *our* people, but not for *white* people. They won't stand for it."
> I stared bleakly into the morning sun. I was nearing my seventeenth birthday and I was wondering if I would ever be free of this plague. What Griggs was saying was true, but it was simply utterly impossible for me to calculate, to scheme, to act, to plot all the time. I would remember to dissemble for short periods, then I would forget and act straight and human again, not with the desire to harm anybody, but merely forgetting the artificial status of race and class. It was the same with whites as with blacks; it was my way with everybody. I sighed, looking at the glittering diamonds in the store window, the rings and the neat rows of golden watches.

Questions

1. What is Griggs trying to teach Wright?

2. Why doesn't Wright want to do what Griggs says?

3. Why does Wright need to do what Griggs says?

4. Where and when do you think this scene takes place? Why?

Answers

1. Griggs is trying to teach Wright how to behave around white people. Specifically, Griggs wants Wright to think before he speaks and acts in front of white people so he behaves in a way that is acceptable to them.

2. Wright doesn't want to "be a slave," to act according to "artificial status of race and class." He doesn't think he should have to change the way he behaves simply because the people around him are of a different color, and he can't remember "to calculate, to scheme, to act, to plot all the time."

3. Wright needs to do what Griggs says in order to eat (in order to be able to get a job to earn money for food). Also, we can infer that he needs to behave properly to avoid getting lynched.

4. We can infer that this scene takes place in the American south some time between the end of the Civil War (1865) and the beginning of the Civil Rights Movement (1960s). Griggs and Wright are free men, but they still have a very low place in society and must constantly cater to the wishes and whims of white people. This indicates a southern, not northern, setting.

JOURNALISM

Newspapers are an important way of keeping ourselves informed. Before television and radio, there were printing presses and newspapers. **Journalism** includes the information and opinions we receive through newspapers, magazines, and other **periodicals** (publications that are issued at regular intervals) as well as television and radio.

The primary purpose of journalism is not to express opinion but to inform readers of **facts**. Newspaper articles, for example, generally report facts about current events: a presidential candidate or election, a catastrophe, a local strike, a new scientific discovery, a new product, etc. Though newspaper and magazine articles may often express an author's opinion, these articles are generally reserved for reviews, which we will discuss in **Commentary**, and for the Opinion-Editorial (Op-Ed) page (these articles are discussed under **Essays**).

One element of journalism is (or should be) **objectivity**. In reporting the facts—the **who, what, when, where, why** and **how**—journalists should (except in editorials and reviews) strive to keep their opinions out of the writing. However, journalists, especially magazine writers, often have a "slant" to their writing. A slant is an approach to the topic that makes it timely and interesting, and sometimes this slant may express the author's stance on the subject.

The following passage is an excerpt from a recent *New York Times* front-page article written by John Kifner. Read it carefully and notice the number of facts given in the first paragraph:

> Eight people in a Harlem clothing store were killed yesterday in a fierce, smoky blaze that the police said they believe was deliberately set as part of a bitter landlord-tenant dispute that led to angry protests in the neighborhood. Among the dead was a man the police suspect set fire to the store before apparently shooting himself in the chest.
>
> The fire gutted the first-floor storefront that housed Freddie's Fashion Mart on Harlem's main thoroughfare, 125th Street, shortly after a tall gunman, waving a .38-caliber handgun, burst into the store across the street from the Apollo Theater. The police said last night that they believed the gunman had once joined a group of demonstrators who had picketed the store in recent weeks in a feud over the threatened eviction of a subtenant, the Record Shack, a neighborhood institution.

Questions

1. What happened?
2. Where did it happen?
3. Who was involved?
4. When did it happen?
5. Why did it happen?
6. How did it happen?

Answers

1. A gunman set fire to a clothing store. Eight people died, including the gunman, who killed himself.

2. It happened in Freddie's Fashion Mart, across from the Apollo Theater on 125th Street in Harlem.

3. A gunman who had demonstrated against the eviction of Record Shack was responsible for the incident. The other people involved are not identified.

4. It had happened the day before the article was written ("Eight people in a Harlem clothing store were killed yesterday…").

5. The incident was apparently sparked by tensions between the landlord (Freddie's Fashion Mart) and sub-tenant (Record Shack). The landlord apparently wanted to evict the sub-tenant, and many angry people were protesting this eviction.

6. The fire was started by the gunman, who then shot himself.

Notice that the author of the article does not include his opinions about the events. He merely presents us with the facts—the who, what, where, when, why, and how. We are left to form our judgments for ourselves.

Journalism serves society in many ways. Most significantly, it keeps us informed about what's happening in our cities and towns, in our country, and around the world.

EVOLUTION OF NON-FICTION

Most of these forms of non-fiction have been evolving for hundreds, if not thousands, of years. The Greeks wrote essays, the Gospels are biographies of Jesus Christ, people have been writing their life stories for centuries, and newspapers have been circulating since the invention of the printing press. A few trends, however, distinguish today's non-fiction. Personal essays, for example, have been enjoying increasing popularity, and biographies are be-

ing written about more "ordinary" people. Many autobiographers are blurring the line between fiction and non-fiction by writing "autobiographical novels."

Journalists, on the other hand, have drifted into a writing style that is more dramatic. Nearly all newspaper articles a decade or two ago provided little more than straight forward facts about the who, what, when, where, why and how of the event. Today, many journalists write not just to inform but to evoke feeling. Here is another front-page article from the *New York Times*, this one written by Robert D. McFadden, about the same Harlem incident. Notice that the facts are the same, but the style is very different:

It was just after 10 o'clock and Harlem had begun like an orchestra to tune up for another great performance. Traffic rumbled. Stores were open. Sidewalk vendors were out. And people moved briskly in the cold December morning, headed for work or shops or favorite haunts, carrying bags like responsibilities.

Across from the famed Apollo Theater, on the southeast corner of 125th Street and Frederick Douglas Boulevard, the gunman appeared out of nowhere at Freddie's Fashion Mart, a white-owned business that for months had been picketed and boycotted over a plan to expand that meant the eviction of a black-owned record shop next door. Feelings had run high. There had been threats.

Questions

1. Did you notice the figurative language and extra detail? List them below.

2. What is the effect of this passage compared to the first one you read? Which one do you like better?

Answers

1. The first instance of figurative language is when Harlem is compared to an orchestra: "Harlem had begun like an orchestra to tune up for another great performance" (simile). Second, the people were "carrying bags like responsibilities" (simile). There is also additional detail in this passage. We know the time ("just after 10 o'clock"), we know that it was a cold day in December, we know the exact intersection, and we know a little more about the tensions that caused the tragedy.

2. Answers may vary, but the main difference to notice is that this passage is more like a story than an article. We do still get the basic facts of the incident, but the facts are presented in a story format, which first establishes the setting and the mood and uses figurative language.

At times this "writing to please" has led to sensationalism, or "writing to sell"—and, of course, the number of tabloids today attests to this. As a reader of newspapers and other periodicals, you must always be ready to consider the integrity of your source.

☞ Practice: Applying Comprehension Strategies.

> **DIRECTIONS:** Read the following passages. Apply the four reading strategies you've learned, as well as your vocabulary skills. Write in the margins and mark up the text as you go. Then answer the questions following each passage.

PASSAGE A—FICTION

The following passage is an excerpt from Edgar Allan Poe's short story "The Tell-Tale Heart."

True!—nervous—very, very dreadfully nervous I had been and am! but why will you say that I am mad? The disease had sharpened my senses—not destroyed—not dulled them. Above all was the sense of hearing acute. I heard all things in the heaven and in the earth. I heard many things in hell. How, then, am I mad? Hearken! and observe how healthily—how calmly I can tell you the whole story.

It is impossible to say how first the idea entered my brain: but once conceived, it haunted me day and night. Object there was none. Passion there was none. I loved the old man. He had never wronged me. He had never given me insult. For his gold I had no desire. I think it was his eye! yes, it was this! One of his eyes whenever it fell upon me, my blood ran cold; and so by degrees—very gradually—I made up my mind to take the life of the old man, and thus rid myself of the eye forever.

Now this is the point. You fancy me mad. Madmen know nothing. But you should have seen me. You should have seen how wisely I proceeded—with what caution—with what foresight—with what dissimilation I went to work!

I was never kinder to the old man than during the whole week before I killed him. And every night, about midnight, I turned the latch of his door and opened it—oh, so gently! And then, when I had made an opening sufficient for my head, I put in a dark lantern, all closed, closed so that no light shone out, and then I thrust in my head. Oh, you would nave laughed to see how cunningly I thrust it in! I moved it slowly—very, very slowly, so that I might lay upon his bed. Ha!—would a madman have been so wise as this? And then, when my head was well in the room, I undid the lantern cautiously—oh, so cautiously—cautiously (for the hinges creaked)—I undid it just so much that a single ray fell upon the vulture eye. And this I did for seven long nights—every night just at midnight—but I found the eye always closed; and so it was impossible to do the work; for it was not the old man who vexed me, but his Evil Eye. And every morning, when the day broke, I went boldly into the chamber, and spoke courageously to him, calling him by name in a hearty tone, and inquiring how he had passed the night. So you see he would have been a very profound old man, indeed, to suspect that every night, just at twelve I looked in upon him while he slept.

1. What do you sense about the narrator from the first paragraph?

2. What do we know about the old man from the statement "For his gold I had no desire."

3. What is it that makes the narrator want to kill the old man?

4. On each of the seven attempts to kill the old man, what does the speaker find?

5. How would you describe the overall tone of the narration?

PASSAGE B—NON-FICTION

This passage is the conclusion of Zora Neale Hurston's "How It Feels to Be Colored Me."

Sometimes, I feel discriminated against, but it does not make me angry. It merely astonishes me. How *can* any deny themselves the pleasure of my company? It's beyond me.

But in the main, I feel like a brown bag of miscellany propped against a wall. Against a wall in company with other bags, white, red and yellow. Pour out the contents, and there is discovered a jumble of small things priceless and worthless. A first-water diamond, an empty spook, bits of broken glass, lengths of string, a key to a door long since crumbled away, a rusty knife-blade, old shoes saved for a road that never was and never will be, a nail bent under the weight of things too heavy for any nail, a dried flower or two still a little fragrant. In your hand is the brown bag. On the ground before you is the jumble it held—so much like the jumble in the bags, could they be emptied, that all might be dumped in a single heap and the bags refilled without altering the content of any greatly. A bit of colored glass more or less would not matter. Perhaps that is how the Great Stuffer of Bags filled them in the first place—who knows?

1. Why is "can" italicized?

2. What do the bags represent? What is the metaphor?

3. Who is the "Great Stuffer of Bags"?

4. What would happen if all the bags were emptied and refilled at random?

5. What does Hurston mean by "A bit of colored glass more or less would not matter"?

Answers

Passage A—Fiction

1. We get the feeling that the narrator is crazy. Although he is trying to convince us otherwise it is apparent that he is quite mad.

2. We know that the old man must be well off.

3. The narrator wants to kill the old man because he is haunted by the old man's evil eye.

4. Each time he tries to kill him he finds that the eye is closed and therefore he cannot kill him.

5. The overall tone of the passage would best be described as anxious or tense.

Passage B—Non-Fiction

1. "Can" is italicized to create a certain tone of disbelief. Hurston is saying she can't understand why people would deny themselves of her company; in fact, she is astonished, so the italics reflect this disbelief and astonishment.

2. The bags are metaphors for our bodies, particularly our skin. Hurston mentions brown, white, red, and yellow bags, the colors typically used when classifying race.

3. The "Great Stuffer of Bags" is God.

4. Hurston suggests that if all the bags were emptied at then refilled at random, they would still be essentially the same as when they began. Though they may have items from other bags now, those items are nearly identical to the original contents.

5. Hurston means that although we may have small differences (the bits of colored glass), they are insignificant when compared to our similarities. Thus, "a bit of colored glass more or less would not matter."

REVIEW

Prose is writing that is not in poetic or dramatic form. Fictional prose is prose that is the product of an author's imagination (invented people and events). Non-fictional prose, on the other hand, tells us about real people and events.

Eight elements of fiction can help us better understand the stories and novels we read:

- plot;
- setting;
- character;
- point of view;
- tone;
- language style;
- symbolism;
- theme.

Plot is the organization of events in a story. The plot is driven by conflict, which is essential to a good story.

Setting locates the story in a particular place and time, thereby providing a specific social and historical context for the action. Setting also helps create the tone or mood of the story. The most important tone to consider is irony—the tension that is created when we know more than one or more of the characters knows.

Characters are the people created by the author to carry the action, language, and ideas of the story. The character who is the "hero" of the story, the one who faces the central conflict, is the protagonist. The antagonist is the person, force, or idea working against the protagonist.

Point of view refers to the narrator of the story. A story may be narrated in the first-, second-, or third-person point of view.

Language and style also help create tone and convey meaning. The author's diction (word choice) is always significant, because each word has a specific denotation, connotation and emotional register. The author's style (the length and complexity of his or her sentences) is also significant because it establishes a rhythm and tone.

Figurative language is an important aspect of language and style. Authors use similes, metaphors, and personification to create effective comparisons and images. Imagery is the representation of sensory experience through language.

Symbols are people or objects invested with additional meaning. They offer authors a chance to layer meaning.

All of these elements add up to the story's theme. The theme of a story is its overall meaning, the ideas it conveys as a whole.

Myths are cultural stories that attempt to explain a practice, belief, or natural phenomenon. Parables are short allegorical tales that illustrate a moral or religious truth. Fables, on the other hand, are brief stories designed to teach a specific moral lesson or reveal a truth about human nature.

There are many types of novels and short stories. They may be divided into four general categories: detective/thriller, adventure/romance, science fiction/fantasy, and social/historical.

Non-fictional texts should be read with these particular elements in mind: voice, language and style, structure, tone, and theme.

In non-fiction, the author is speaking to us directly, so the voice he or she uses is his or her own—it is not filtered through a narrator. The tone of this voice is created by the point of view and the language and style. How intimate is the author? Does he or she speak to us using "I," "you," or "he"? What type of words does he or she use—long and technical, or short and colloquial?

The way an author structures a non-fictional text is always revealing. We should notice how an essay is divided and organized. Then we should ask why it is structured and organized in that way.

Essays are non-fictional texts that are essentially arguments: they make an assertion about their subject and then support it. The key to understanding an essay is being able to determine its thesis, the author's position on the subject. Then, we must examine how the essayist supports that assertion.

Sometimes the thesis is not clearly expressed in a thesis statement, and we need to look for clues as to the main idea. This main idea must encompass the whole essay, not just some of its parts, and it is often suggested by the title.

There are many types of essays called rhetorical modes, such as description, narration, persuasion, etc. But all essays are essentially arguments. A satire is a text that exposes and ridicules its subject. Satire uses verbal irony, hyperbole, parody, and other techniques to prove its point.

A biography is the true account of a person's life. An autobiography is also a true account of a person's life, but it is told by that person him or herself rather than by an outside biographer. In both cases, the writers will describe the people, places and events that shaped the lives of their subjects.

A biography is usually quite objective since many sources provided information about the subject. An autobiography, on the other hand, is much more subjective, since the writer is telling his or her own story. At the same time, however, an autobiography is usually much more intimate and revealing.

Journalism is writing that informs readers of facts about events taking place in our world. Journalists often write opinion pieces, but these are editorials and should be considered essays. Reviews, on the other hand, are commentary. Otherwise, journalists strive for objectivity in their articles as they describe the who, what, when, where, why, and how regarding the event.

Interpreting Literature
and the Arts

Reading Poetry

INTERPRETING LITERATURE AND THE ARTS

READING POETRY

WHAT IS POETRY?

Though most of us can recognize **poetry** when we see it, we may have a difficult time defining it. Is poetry words that rhyme? Words that sound good together? A group of short lines? Yes, and no, and much more.

What makes poetry both so difficult and so beautiful is that, unlike an essay or a short story, a poem has to say a great deal in a very limited space and in a restricted form. Therefore, each word must be chosen with extreme care. A poem must say what it has to say in very few words. That's how a poem gets its intensity: Every word counts.

Unlike prose, in which we are captivated primarily by the story line and/or the author's argument, in poetry we are captivated primarily by the words themselves and the images, emotions, and experiences they elicit. Poetry is not only the story of an experience or an emotion—it is also the story of the relationship of the words that express this experience or emotion.

Poetry also differs from prose in that it is meant to be *heard* as well as *read*. It gets its meaning not just from *what* it says but from the sound of the words chosen to say it. Thus, to understand a poem, we must look at both its **sound** and its **sense**. Ideally, as you learn more about poetry, you will begin to delight in the language and layers of meaning a poem offers.

ELEMENTS OF SOUND

Rhyme is the repetition of an identical or similar stressed sound or sounds at the end of a word. Rhyme essentially does two things: one, it suggests order; and two, it suggests a relationship between the rhymed words.

There are several types of rhyme: an **exact rhyme**, in which the last syllables (last consonant and vowel combination) rhyme (as in sugg**est** and requ**est**); a **half-rhyme**, in which only the final consonants "rhyme" (as in ca**t** and ho**t**); or an **eye rhyme**, which simply looks like a rhyme (the word endings are spelled the same) but it isn't (the words don't sound the same, as in **cough** and **bough**).

Another element of sound—one often employed in prose as well—is **alliteration**. Alliteration is simply the repetition of sounds. It is usually at the beginning of words, but it can also be at the end of words, throughout the words, or all three. For example, look at these pairs of words: **fail** and **feel**; **rough** and **roof**; **pitter** and **patter**. Now look at the following excerpt from a poem by A.C. Swinburne, "When the Hounds of Spring." See if you can circle all the instances of alliteration:

> For winter's rains and ruins are over,
>   And all the seasons of snows and sins;
> The days dividing lover and lover,
>   The light that loses, the night that wins.

Did you notice the repetition of **r** in the first line, particularly in the words *rains* and *ruins*, which differ by only one letter? Did you notice the repetition of the **s** sound in the second line? And **d** in the third, **l** in the last? Did you notice the **internal rhyme** in the last line as well (*light* and *night*)?

Writers of all kinds of literature use alliteration because when sounds are repeated, they create certain effects. Soft sounds, like *a, e, i, o, u, l, r, m, n, w,* and *y* create a pleasant, musical effect, whereas harder sounds, such as the consonants *b, d, k, p, t,* and *g* are more explosive, harder on the ear and less pleasant. Therefore, authors—especially poets—will use alliteration of sounds that create the right mood or reflect the theme of their text.

Onomatopoeia, another element of sound, occurs when the sound of a word echoes or suggests its meaning. The meaning *is* the word—like *hiss, buzz, moan, murmur,* and *quack,* for example. Look at the following excerpt from Robert Frost's "Out, Out—" and see if you can identify the onomatopoeia:

OUT, OUT—

(1) The buzz saw snarled and rattled in the yard
(2) And made dust and dropped stove-length sticks of wood.
(3) Sweet-scented stuff when the breeze drew upon it.
(4) And from there those that lifted eyes could count
(5) Five mountain ranges one behind the other
(6) Under the sunset far into Vermont.
(7) And the saw snarled and rattled, snarled and rattled,
(8) As it ran light, or had to bear a load.

Did you correctly identify "buzz," "snarled," and "rattled" in lines 1 and 7?

With onomatopoeia, as you can see, sound and sense literally become one, and the sound of the poem comes more vibrantly to life.

Meter refers to both the number of syllables in a line and the stress of those syllables. For example, in **iambic pentameter**, the most common metrical pattern, stresses fall on the second syllable of words (or every other syllable), like re*fine* and di*vine*. This pattern of stressed and unstressed syllables is called *iambic*. Each line in iambic pentameter form has five ("penta") feet. A foot, in iambic form, is the term for two syllables, one stressed and one unstressed. So there are ten syllables on each line in iambic pentameter form, and five feet.

Read "The Eagle," by Alfred, Lord Tennyson, aloud to hear how the metrical pattern works. Then, determine the metrical pattern. (Note: When you read poetry, don't stop at the end of a line unless punctuation tells you do to so. If no punctuation stops you, keep moving right into the next line. Likewise, pause whenever there is punctuation within a line.)

THE EAGLE

(1) He clasps the crag with crooked hands;
(2) Close to the sun in lonely lands,
(3) Ringed with the azure world, he stands.

(4) The wrinkled sea beneath him crawls;
(5) He watches from his mountain walls,
(6) And like a thunderbolt he falls.

Questions

1. What is the metrical pattern in "The Eagle"?

2. What is the rhyme scheme?

3. Can you find two examples of personification?

4. Can you find the simile?

5. Did you notice any alliteration? Where?

Answers

1. The metrical pattern is iambic tetrameter. The stress falls on every other syllable ("He *clasps* the *crag* with *crook*ed *hands*") and there are four (tetra) feet in each line.

2. The rhyme scheme is two triplets: aaa, bbb.

3. Two examples of personification are 1) that the eagle has hands and 2) that the sea crawls.

4. The simile is "*like* a thunderbolt he falls," comparing the eagle's fall to a thunderbolt.

5. There is alliteration in line 1 with *clasps*, *crag*, and *crooked*. There is also alliteration in line 2 with *lonely lands*.

All of these elements of sound, combined with the poem's diction and **structure,** which we will discuss shortly, add up to the **rhythm** of the poem. The rhythm is the overall effect of a poem's sound. Does it, for example, roll easily off your tongue? Or is it slow and heavy? Does it sound mesmerizing because of its repetition? What words are stressed or stand out? Does it sound like classical music, or more like a heavy metal concert? Like the banging of drums, or the singing of birds? The rhythm will reflect both the poem's mood and its theme.

ELEMENTS OF SENSE

Essentially, the elements that help us make sense of poetry are the same as those that help us make sense of fiction: **plot, setting, point of view, tone, language and style** (*especially* language and style), **symbolism,** and **theme.** There are a few differences, however.

One difference you should be aware of is that in poetry, the person who is speaking—the **persona**—is often unclear or unnamed. We must use the clues provided by the poem to determine whose voice is speaking. (Remember that the persona is *not* the poet, just as in a story the narrator is not the author.) The persona is essentially a combination of point of view and character. It is the voice that the poet has chosen for the poem, and though it may be very autobiographical, the voice is a fictional (imaginary) "character" who shares a story, image, or experience.

For example, read Katharyn Howd Machan's "Hazel Tells LaVerne." When you've finished, summarize the poem and write down your observations. Then see how much you can determine about the speaker.

HAZEL TELLS LAVERNE

(1) last night
(2) im cleanin out my
(3) howard johnsons ladies room
(4) when all of a sudden
(5) up pops this frog
(6) musta come from the sewer
(7) swimmin aroun an tryin ta
(8) climb up the sida the bowl
(9) so i goes ta flushm down
(10) but sohelpmegod he starts talkin
(11) bout a golden ball
(12) an how i can be a princess
(13) me a princess
(14) well my mouth drops
(15) all the way to the floor
(16) an he says
(17) kiss me just kiss me
(18) once on the nose
(19) well i screams
(20) ya little green pervert
(21) an i hitsm with my mop
(22) an has ta flush
(23) the toilet down three times
(24) me
(25) a princess

First, *summarize* the action in the poem. What happens?

_____

_____

_____

Now, look closely at the language and structure of the poem. What did you notice?

I noticed:

1. _____

2. _____

3. _____

4. _____

5. _____

Your *summary* might sound something like this:

> While Hazel was cleaning the bathroom at a Howard Johnson's, a frog came up out of the toilet and told her she could be a princess if she kissed him on the nose. She screamed at him, hit him with her mop, and flushed him down the toilet.

Did you *notice*:

- That there is no capitalization or punctuation?

- That there are several instances of words that run together and many word endings are dropped, as in casual speech?

- That "me a princess" is repeated twice?

- That she is possessive about the ladies' room ("my")?

- That the talking frog is from a fairy tale?

Now, what can you tell about the speaker, Hazel? What does she do? How does she feel about what she does? What can you tell about her life and personality based on how she speaks? Most importantly, how do you think she feels about herself? Why?

If you read carefully, you should be able to tell that Hazel is a cleaning woman for a Howard Johnsons (lines 2–3) and that she is proud of what she does since she uses the possessive "my howard johnsons ladies room" in lines 2–3. The lack of punctuation and capitalization, as well as the non-standard English (such as "I screams" in line 19) indicate that she has not had much formal education.

The way that Hazel reacts to the frog—screaming at him and hitting him with her mop—indicates that she doesn't believe in fairy tales, like the one suggested by the poem. (In that fairy tale, a frog tells a lonely princess that he'll turn into a prince if she kisses him. She does, and the frog—who is really a prince under a spell—turns back into a prince and marries her.) Perhaps Hazel simply doesn't believe that *she* could possibly be a princess, that a fairy tale ending could come true for someone like her. This is suggested by the fact that she repeats "me a princess" twice.

We can also see from what Hazel says to the frog ("ya little green pervert") that she is quick to believe that the frog is trying to get something from her rather than give her something. And Hazel doesn't want to risk being a fool enough to believe the frog *might* be telling the truth.

In poetry, we must also pay particular attention to **language and style**. In a poem, every word counts—so the words a poet uses must be chosen with special care. As in prose, the words a poet chooses convey the poet's attitude towards the subject of the poem. In addition, some words work on many levels of meaning. They may have several denotations, connotations, and suggestions.

Look carefully at the following poem by Stephen Crane, for example. Notice his choice of words (**diction**):

WAR IS KIND

(1) Do not weep, maiden, for war is kind.
(2) Because your lover threw wild hands toward the sky
(3) And the affrighted steed ran on alone,
(4) Do not weep.
(5) War is kind.
(6)   Hoarse, booming drums of the regiment
(7)   Little souls who thirst for fight,
(8)   These men were born to drill and die
(9)   The unexplained glory flies above them
(10)   Great is the battle-god, great, and his kingdom—
(11)   A field where a thousand corpses lie.
(12) Do not weep, babe, for war is kind.
(13) Because your father tumbled in the yellow trenches,
(14) Raged at his breast, gulped and died,
(15) Do not weep.
(16) War is kind.
(17)   Swift, blazing flag of the regiment
(18)   Eagle with crest of red and gold,
(19)   These men were born to drill and die
(20)   Point for them the virtue of slaughter
(21)   Make plain to them the excellence of killing
(22)   And a field where a thousand corpses lie.
(23) Mother whose heart hung humble as a button
(24) On the bright splendid shroud of your son,
(25) Do not weep.
(26) War is kind.

Questions

1. What is the subject of the poem?

2. To what three people are the parts of this poem addressed?

3. What repetition did you notice?

Answers

1. The subject is war.

2. The first part of the poem is addressed to a maiden, the second to a child, and the third to a mother. All three have lost loved ones (a lover, a father, a son) in a war.

3. At first glance, the repetition of "Do not weep./ War is kind" makes it seem as if that is the speaker's message to these listeners. But careful readers can tell that the speaker does *not* feel that war is kind; in fact, he feels the opposite. We can tell this from the other repeated lines and the diction.

First, look at the other repeated lines: "These men were born to drill and die" and "A field where a thousand corpses lie." The first makes the men sound as if they are simply war machines; that they had no other purpose in life than to die for their country. But the people being addressed here—the lover, child, and mother—know that there is much more to these men who have died in the war. The second of these lines is significant in its diction. Crane uses the word "corpses" to emphasize the loss of life. He could say "A field where a thousand soldiers lie" or "A field where a thousand enemies lie"—but he specifically says "corpses."

The diction in line 10 is also significant. The battle-god and his kingdom are "great," and a homonym of *great* is *grate*, which means to shred into small pieces, or to have an unpleasant or irritating effect. Men in battle are, unfortunately, shred by bullets and shrapnel; and to say war is unpleasant or irritating is certainly an understatement.

Language is especially telling in lines 20–21. "Point for them the *virtue* of *slaughter*," line 20, clearly suggests that the speaker does not feel war is virtuous. How can we tell? Because if he did feel war was virtuous, he would not have used the word "slaughter." It is much too graphic and violent. Instead, he would have used a more glorifying term such as "defense."

Similarly, "Make plain to them the *excellence* of *killing*" (line 21) juxtaposes two terms that should not go together. "Excellence," like "virtue," is a positive term; but it is matched with a word as heavy as "killing." Instead of making plain the excellence of war, the speaker is making plain the horror of it.

In both of these lines we have what is called a **paradox**: a statement or phrase that seems to contradict itself or to conflict with common sense but which contains some truth. These two pairs of words seem to contradict each other—seem not to make sense together—especially in the eyes of the maiden whose lover has died, the child whose father has died, the mother whose son has died. However, for those in command of a war, there *is* excellence in killing; there *is* virtue in slaughter—of the enemy. The **irony** of this paradox is that, through "excellence of killing," their own men have also been slaughtered.

The ultimate paradox is, of course, the much repeated "War is kind." We must ask "To whom?" To the heroes who are glorified after a war is over? Perhaps. Ultimately, it seems that Crane may have played quite a trick by using the word "kind." To say you are giving something "in kind" is to give something other than money in return for some "good" that has been exchanged. Here, the gift "in kind"—in return for the soldiers' services—is death.

Imagery is another important aspect of language to consider when reading poetry. Much of poetry is an attempt to engage the reader's senses, to get the reader to see, hear, smell, and feel something that the speaker is describing. This is imagery. Some poems have this development of imagery as their sole purpose. Consider the following poems, for example, both from the year 1916:

IN A STATION OF THE METRO*

The apparition of these faces in the crowd;
Petals on a wet, black bough.

(*Paris subway)   —Ezra Pound

FOG

(1)   The fog comes
(2)   on little cat feet.

(3)   It sits looking
(4)   over harbor and city
(5)   on silent haunches
(6)   and then moves on.

—Carl Sandburg

In the first of these poems, you are asked by the title to imagine yourself in a Paris subway station. What is such a station like? Crowded with empty white faces, blurry faces as people rush anonymously to their destinations. Now, picture a tree, a dogwood perhaps, in the spring, with its white flowers in bloom, immediately after a rainstorm. Petals from the flowers would have fallen onto the thick water-darkened lower branches. This is the picture Ezra Pound wants us to compare to the Metro. The station, with its dark tunnels, is the bough; the nameless white faces of the Parisians rushing through it are the petals. It is like the snapshot of a camera that is out of focus so no facial features are distinguishable.

"Fog," on the other hand, uses personification to create its powerful imagery. We are asked to imagine fog coming in like a cat—softly, lightly, stealthily. Like a cat, it sits silently on its haunches, surveys, and then "moves on." It is a very vivid picture.

In both poems, the poets are asking you to see their subjects in a new way. Have you ever thought of fog as a cat before? Probably not. But isn't it a very accurate and clever comparison? And have you ever thought of people in a subway station as petals, or apparitions (ghosts)? We might ask "Who are they?" Where are they going? Where are they from? The poem can take off in the reader's mind in endless directions.

This, then, is the power of imagery. It enables us to imagine things in a new, vivid way through the eyes, ears, nose, taste, or skin of another.

Finally, in poetry, one element that is essential to meaning is form. Thus, when reading poetry, we must take a close look at a poem's structure.

POETIC FORMS: THE IMPORTANCE OF STRUCTURE

A new element to consider in poetry is **structure**. While the structure of a story is important, the story's structure is largely dictated by its plot—the arrangement of events. But in a poem, plot is usually very limited. In fact, in many, if not most, poems, there is very little action. When this is the case, we must focus more on the **subject** than the plot. We must ask ourselves "what is it about?" rather than "what happens?" We may ask, "Is the poem describing a person? A place? An event? A thought process? A moment of truth? A painful experience? A beautiful vision?" Then we must look carefully at the poem's structure, which often dictates the poet's language and style.

When we consider structure, we must consider the overall arrangement of the poem. A **narrative** poem, like "Hazel Tells LaVerne," tells a story, and, like fiction, it is driven forward by action. An **argument** poem, like "War Is Kind," attempts to convince the reader and/or the person being addressed by the poem of something—some truth, a certain way of seeing things. An **imagistic** poem, such as "Fog," seeks mainly to create a vivid image. There are also **visual** poems, which use their form to create a picture of their subject, thereby reinforcing or reflecting the poem's ideas. For example, look at George Herbert's "Easter Wings" (1633):

EASTER WINGS

(1) Lord, who createdst man in wealth and store,
(2)   Though foolishly he lost the same,
(3)   Decaying more and more,
(4)   Till he became
(5)   Most poor:
(6)   With thee
(7)   O let me rise
(8)   As larks, harmoniously,
(9)   And sing this day thy victories:
(10) Then shall the fall further the flight in me.

(11) My tender age in sorrow did begin:
(12)   And still with sicknesses and shame
(13)   Thou didst so punish sin,
(14)   That I became
(15)   Most thin.
(16)   With thee
(17)   Let me combine,
(18)   And feel this day thy victory;
(19)   For, if I imp my wing on thine,
(20) Affliction shall advance the flight in me.

What does the poem look like because of how the lines are arranged? Does it look like two hourglasses, perhaps? Now look at it sideways. What does it look like? Do you see a pair of butterflies? How about two pairs of angel wings? This is what the form is meant to portray, because you can see from the title of the poem and from its content that it is a very religious poem.

Notice the thinnest points of the poem. What do those lines say? "Most poor:/ With thee" and "Most thin./ With thee." Can you see here how the literal meaning of these lines is reflected in the physical shape of the poem?

Two other structural arrangements remain. In addition to narrative, argumentative, imagistic, and visual poetry, there are also **elegies** and **odes**. An elegy is a lament for the loss or death of someone or something, and it is usually a mournful poem. An ode, on the other hand, is usually celebratory; it celebrates a person, place, thing, or event and expresses profound thoughts about its subject.

John Keats' "Ode on a Grecian Urn" (1819), for example, is an ode that turns a simple vase into a celebration of history. In the ode, the vase becomes a source of wisdom that teaches the speaker about the timelessness of art and the earthly "truth" of beauty. The last lines of the poem read:

> "Beauty is truth, truth beauty,"—that is all
> Ye know on earth, and all ye need to know.

By structure we also mean how the poem is physically organized. How many lines are there? How long are they? What is the rhyme scheme, if any? What is the metrical pattern? Before we go any further, let us distinguish between three general types of poems with varying degrees of structure: **rhymed verse, blank verse,** and **free verse.**

Rhymed and Metered Verse

Of all poems, those written in **rhymed and metered verse** are the most restricted by structure. **Diction** in particular is controlled by the rhyme scheme and metrical pattern of this type of poem. The words the poet chooses must convey just the right meaning and still have the right ending to fit the rhyme scheme, the right number of syllables to fit the metrical pattern, and the right stresses to fit in the right place in the line.

Three types of rhymed and metered verse are the **sonnet, ballad,** and **villanelle**. In each of these poems, poets have to follow both a rhyme scheme and metrical pattern. A sonnet, for example, is composed of fourteen lines and almost always written in iambic pentameter (if not, the poet will usually substitute a different metrical pattern). In the Italian, or Petrarchian, sonnet, the lines are divided into a **stanza** (group of lines working together, like a paragraph in an essay) of eight lines and a stanza of six with the following rhyme scheme: *abba abba cdcdcd* (or *cdecde* or *cdccdc*).

In the Shakespearean sonnet, on the other hand, the fourteen lines are separated into three **quatrains** (groups of four lines) and a **couplet** (a pair of lines) with this rhyme scheme: *abab cdcd efef gg*. The closing couplet usually "solves" the problem presented in the quatrains.

A **ballad** is essentially a story-song. In its most formal form, it is composed of quatrains that rhyme *abcb*. One popular form has quatrains with eight, six, eight, and six syllables per line. This pattern is then repeated over and over. There are, of course, many variations of this form. Ballads usually tell a single story with emphasis on the action rather than description or emotion. Most older ballads told tragic tales, particularly stories about tragic love affairs. The first lines of the old ballad "The Unquiet Grave," spoken by a man above his lover's grave, are a perfect example:

THE UNQUIET GRAVE

> The wind doth blow today, my love,
>   And a few small drops of rain:
> I never had but one true-love,
>   In cold grave she was lain:

A **villanelle** is perhaps the most difficult of these rhymed and metered poems. In a villanelle, there are five stanzas, each of three lines that rhyme *aba* and a final quatrain that rhymes *abaa*. In addition, there are only two rhymes in the poem. Furthermore, line one must be repeated in lines six, 12, and 18; line three must be repeated in lines nine, 15, and 19. Dylan Thomas wrote perhaps today's most well-known villanelle, "Do Not Go Gentle Into That Good Night":

DO NOT GO GENTLE INTO THAT GOOD NIGHT

(1) Do not go gentle into that good night,
(2) Old age should burn and rave at close of day;
(3) Rage, rage against the dying of the light.

(4) Though wise men at their end know dark is right,
(5) Because their words had forked no lightning they
(6) Do not go gentle into that good night.

(7) Good men, the last wave by, crying how bright
(8) Their frail deeds might have danced in a green bay,
(9) Rage, rage against the dying of the light.

(10) Wild men who caught and sang the sun in flight,
(11) And learn, too late, they grieved it on its way,
(12) Do not go gentle into that good night.

(13) Grave men, near death, who see with blinding sight
(14) Blind eyes could blaze like meteors and be gay,
(15) Rage, rage against the dying of the light.

(16) And you, my father, there on the sad height,
(17) Curse, bless, me now with your fierce tears, I pray,
(18) Do not go gentle into that good night.
(19) Rage, rage against the dying of the light.

Questions

1. Can you write out the rhyme scheme?

2. What is the metrical pattern?

3. Who is the persona speaking to?

4. What is the persona's message to this person?

5. Did you notice the contrast between day and night? Between dark and light? What might these symbolize? What is the "dying of the light"?

6. Did you notice any alliteration?

7. What is the significance of the adjective used to describe men in line 13 ("grave")?

Answers

1. The rhyme scheme is *aba aba aba aba aba abaa.*

2. The metrical pattern is iambic pentameter. The stress falls on the second syllable of words (or every other syllable) and there are five feet per line.

3. The persona is speaking to his father.

4. The persona's message to his father is to fight death ("the dying of the light").

5. The "dying of the light" is death; light is life, darkness is death.

6. There is much alliteration in this poem, largely because of the repetition of the first and

third lines. In the first line, "Do not go gentle into that good night" repeats the *g* and *n* sounds. The third line repeats the *g* sound in "Rage, rage against the dying of the light." In line 14, there is also repetition of the *bl* sound in "blind" and "blaze."

7. The adjective "grave" is significant because these "grave men, near death" will soon be lying in their graves. Thomas is making a play on the word "grave." As an adjective, "grave" means important, serious, or solemn, but we are clearly supposed to think of the noun "grave" as well.

Blank Verse

Blank verse is less restricted because it is guided only by meter and not by rhyme. A **haiku**, for example, is a short imagistic poem composed of three unrhymed lines with words that total 17 syllables. Line one has five syllables, line two has seven, and line three has five. The haiku is very popular in Japan, where it originated as a game. Many haiku lose their metrical pattern in translation, but the haiku remains a beautiful and popular form because of its brevity and focus on a single image, as in the haiku below:

THE FALLING FLOWER

What I thought to be
Flowers soaring to their boughs
Were bright butterflies.
—Moritake [1452–1540]

Here is a more contemporary haiku written by W.S. Merwin:

SEPARATION

Your absence has gone through me
Like thread through a needle.
Everything I do is stitched with its color.

You'll notice that this poem doesn't quite match the metrical pattern of a haiku (there are 24 syllables here), but it has the three line arrangement and is powerfully imagistic.

Try to compose your own haiku. Remember, you want to create a powerful sensory image in 17 syllables (five, seven, and five syllables per line).

Free Verse

Free verse is literally that: poetry free from any restrictions of meter and rhyme. That does not mean, however, that there is no rhyme or reason to a free-verse poem. In fact, there almost always is—but it is a rhyme and reason imposed by the poet rather than by traditional poetic forms. For example, in a free-verse poem, the first stanza may have four lines; the second, three lines; the third, two; and the fourth, one.

Rather than fitting a rhythmical or metrical pattern, many free-verse poems fit a thematic or repetitive pattern. Kenneth Fearing's "Ad" is an excellent example of a thematically structured free-verse poem. Read it carefully, noticing as much as you can about the elements of sound and sense, and see if you can explain the structure of the poem (why it is arranged the way it is):

AD

(1) *Wanted:* Men;
(2) Millions of men are *wanted at once* in a big new field;
(3) *New, tremendous, thrilling, great.*
(4) If you've ever been a figure in the chamber of horrors,
(5) If you've ever escaped from a psychiatric ward,
(6) If you thrill at the thought of throwing poison into wells, have heavenly visions
(7) of people, by the thousands, dying in flames—

(8) *You are the very man we want*
(9) We mean business and our business is *you*
(10) *Wanted:* A race of brand-new men.

(11) Apply: Middle Europe;
(12) No skill needed;
(13) No ambition required; no brains wanted and no character allowed;

(14) *Take a permanent job in the coming profession*
(15) Wages: *Death.*

Questions

1. Can you explain the structure of the poem? What does it remind you of?

2. What type of person is the "persona" of the poem looking for? Why?

3. What can you infer about the poet's attitude toward war? Support your answer with specific references to words and lines from the poem.

4. How do you think the poet feels about how governments recruit soldiers for war? How can you tell?

Answers

1. You should have noticed, of course, that the poem is structured like a detailed help-wanted ad and that the writer of the ad is seeking men to fight in a war—soldiers.

2. The "persona" of the poem is looking for men who are evil looking (line four), crazy (line five), and criminal (line six). The ad calls for a person who has been "a figure in the chamber of horrors" (scary looking), someone who has "escaped from a psychiatric ward" (crazy), someone who would "thrill at the thought of throwing poison into wells" (evil) and who has "heavenly visions / of people, by the thousands, dying in flames—."

3. This question asks you about the poem's theme. We can infer from the type of person the "ad" is recruiting that the poet despises war. These recruits need no skill nor ambition, and if they have brains (intelligence), they are not wanted. If they have character, they are simply not allowed. Thus, the poet feels that the people who fight in wars have no character, no brains, no skill; that they are horrific people who are thrilled by death and destruction.

4. By writing this poem in the form of an ad, the poet is clearly commenting on the way that governments recruit soldiers. The ad is filled with glamorous and exciting terms such as *"New, tremendous, thrilling, great" (line* 3) and "You *are* the very man we want" (line 8). The ad also exaggerates the newness of the "field" (even though war is as old as man) by calling the war "a big new field" (line 2), calling for a "race of brand-new men" (line 10), and calling fighting "*the coming* profession" (line 14). Even the "wages"—death—is italicized to make it seem exciting.

These last questions have essentially asked you about the poem's theme—the poem's message. Was it clear to you that the poet despises war?

On the other hand, there is free verse that is very restricted, not by meter or rhyme, but by other constraints. A **sestina**, for example, is a poem of 39 lines of any length and metrical pattern that usually doesn't rhyme. Instead, it is held together by repetition. A sestina is divided into stanzas of six lines each. The end-words of the first stanza must be used as the end-words in every stanza but in a different (but predetermined) order. The final three lines must include all six of the first stanza's end-words. It is a very difficult task. Perhaps the most well-known sestina is Elizabeth Bishop's "Sestina." Read it carefully and answer the questions that follow.

SESTINA

(1) September rain falls on the house.
(2) In the failing light, the old grandmother
(3) sits in the kitchen with the child
(4) beside the Little Marvel Stove,
(5) reading the jokes from the almanac,
(6) laughing and talking to hide her tears.

(7) She thinks that her equinoctial tears
(8) and the rain that beats on the roof of the house
(9) were both foretold by the almanac,
(10) but only known to a grandmother.
(11) The iron kettle sings on the stove.
(12) She cuts some bread and says to the child,

(13) *It's time for tea now*; but the child
(14) is watching the teakettle's small hard tears
(15) dance like mad on the hot black stove,
(16) the way the rain must dance on the house.
(17) Tidying up, the old grandmother
(18) hangs up the clever almanac

(19) on its string. Birdlike, the almanac
(20) hovers half open above the child,
(21) hovers above the old grandmother
(22) and her teacup full of dark brown tears.
(23) She shivers and says she thinks the house
(24) feels chilly, and puts more wood in the stove.

(25) *It was to be*, says the Marvel Stove.
(26) *I know what I know*, says the almanac.
(27) With crayons the child draws a rigid
(28) house
(29) and a winding pathway. Then the child
(30) puts in a man with buttons like tears
and shows it proudly to the grandmother.

(31) But secretly, while the grandmother
(32) busies herself about the stove,
(33) the little moons fall down like tears
(34) from between the pages of the almanac
(35) into the flower bed the child
(36) has carefully placed in the front of the house.

(37) *Time to plant tears*, says the almanac.
(38) The grandmother sings to the marvelous stove
(39) and the child draws another inscrutable house.

Questions

1. Notice the six words that end the lines in each stanza. What do they have in common? What relationships can you draw between these words?

2. What other words are repeated often?

3. What figurative language did you notice?

Answers

1. There are several relationships that can be drawn between the six words that end the lines in each stanza—*house, grandmother, child, stove, almanac,* and *tears*. First, most of the words are domestic, having to do with the house and family. Second, there is the obvious relationship between *child* and *grandmother,* who represent the past and future of the family in this house. The *almanac* is a very old resource for predicting the weather and other occurrences, so it, too, links the present and the past. *Tears* is the hardest word to connect; its source is a mystery in the poem. But the almanac's advice is to plant the tears, so the suggestion is to put old sorrows and pains away into the soil so that they will grow into something beautiful.

2. The two other words that are repeated often are *sings* and *rain*.

3. There is personification in line 11 when the "iron kettle sings on the stove"; the drops of perspired steam on the kettle are "small hard tears" (metaphor, line 14); the tears "dance like mad" (personification, line 15); the man the child draws has "buttons like tears" (simile, line 29); and the moons fall from the almanac "like tears" (simile, line 33).

MAJOR MOVEMENTS IN POETRY

Essentially all English literature began as poetry. Plays were written in verse, and stories were either **ballads** or **epics**, stories about heroic deeds usually told in rhyming couplets. This is partly because literature preceded the printing press, and it was often too difficult or too expensive to get handwritten copies of texts. Instead, **bards**, as they were called, shared stories orally, and rhyming made it much easier for bards to memorize their tales.

In Medieval times (500–1400), poetry and drama began to break off into separate genres. Early poetry was often narrative. Geoffrey Chaucer's *Canterbury Tales* is a good example, though epics like *Beowulf* were still popular.

During the Renaissance (1400–1700), poets began experimenting with structure, and several poetic forms were "born." The **sonnet** is the most important of these.

The Romantic poets of the late-eighteenth and nineteenth centuries began to write poetry about everyday people and things, particularly emotions. They often wrote of emotional extremes and emphasized the wonders of man's imagination. **Odes** and **elegies** were popular forms at this time.

Twentieth-century poets have generally preferred free verse to other forms of poetry. This is partly a result of World War I. After the war, writers began to feel that the world was not as rigidly structured (socially, politically, morally) as they had once thought it to be, that old ways of thinking had to be revised as they had led to the death of millions. Poets, then, wanted their work to reflect new ways of thinking about and understanding the world.

Other literary traditions, of course, have their own histories. The majority of the texts you will be tested on, however, come from English and American writers.

Pre-GED Interpreting Literature and the Arts

☞ Practice: Applying Comprehension Strategies

DIRECTIONS: Read the following passages. Apply the four reading strategies you've learned as well as your vocabulary skills. Write in the margins and mark up the text as you go. Then answer the questions following each passage.

POEM A

The following poem, "Dulce Et Decorum Est," was written by Wilfred Owen. The Latin sentence, *Dulce et decorum est pro patria mori,* found in the last two lines, means "It is sweet and fitting to die for one's country."

DULCE ET DECORUM EST

(1) Bent double, like old beggars under sacks,
(2) Knock-kneed, coughing like hags, we cursed through sludge,
(3) Till on the haunting flares we turned our backs
(4) And towards our distant rest began to trudge.
(5) Men marched asleep. Many had lost their boots
(6) But limped on, blood-shod. All went lame; all blind;
(7) Drunk with fatigue; deaf even to the hoots
(8) Of tired, outstripped Five-Nines that dropped behind.

(9) Gas! GAS! Quick, boys!—An ecstasy of fumbling,
(10) Fitting the clumsy helmets just in time;
(11) But someone still was yelling out and stumbling,
(12) And flound'ring like a man in fire or lime...

(13) Dim, through the misty panes and thick green light,
(14) As under a green sea, I saw him drowning.

(15) In all my dreams, before my helpless sight,
(16) He plunges at me, guttering, choking, drowning.

(17) If in some smothering dreams you too could pace
(18) Behind the wagon that we flung him in,
(19) And watch the white eyes writhing in his face,
(20) His hanging face, like a devil's sick of sin;
(21) If you could hear, at every jolt, the blood
(22) Come gargling from the froth-corrupted lungs,
(23) Obscene as cancer, bitter as the cud
(24) Of vile, incurable sores on innocent tongues,—
(25) My friend, you would not tell with such high zest
(26) To children ardent for some desperate glory,
(27) The old Lie: Dulce et decorum est
(28) Pro patria mori.

1. What is the "old Lie"?

2. The person being addressed by the speaker is

   (1) a soldier.
   (2) a civilian.
   (3) a reporter.
   (4) the enemy.

3. How does the soldier die?

4. List several instances of figurative language.

5. What is the rhyme scheme?

6. What is the metrical pattern?

7. What is the theme of this poem?

POEM B

The following poem is John Donne's "Death Be Not Proud," published about 1610:

DEATH BE NOT PROUD

(1) Death be not proud, though some have called thee
(2) Mighty and dreadful, for thou art not so;
(3) For those whom thou think'st thou dost overthrow
(4) Die not, poor death, nor yet canst thou kill me.
(5) From rest and sleep, which but thy pictures be,
(6) Much pleasure, then from thee much more must flow,
(7) And soonest our best men with thee do go,
(8) Rest of their bones, and soul's delivery.
(9) Thou art slave to fate, chance, kings, and desperate men,
(10) And dost with poison, war, and sickness dwell,
(11) And poppy, or charms can make us sleep as well,
(12) And better than thy stroke; why swell'st thou then?
(13) One short sleep past, we wake eternally,
(14) And death shall be no more; Death, thou shalt die.

1. To whom is the persona speaking?

2. To what does Donne compare Death?
   (1) Bones
   (2) Sleep
   (3) Fate
   (4) Charms

3. To what is Death a slave? Can you explain this line?

4. The end result of death, according to this poem, is that "we wake eternally" and "death shall be no more." Why? What is the "eternal wake"?

5. This poem is
   (1) a ballad.
   (2) a villanelle.
   (3) an Italian sonnet.
   (4) a Shakespearean sonnet.

POEM C

The following is a poem by contemporary poet Marge Piercy:

THE SECRETARY CHANT

(1) My hips are a desk.
(2) From my ears hang
(3) chains of paper clips.
(4) Rubber bands form my hair.
(5) My breasts are wells of mimeograph ink.
(6) My feet bear casters.
(7) Buzz. Click.
(8) My head
(9) is a badly organized file.
(10) My head is a switchboard
(11) where crossed lines crackle.
(12) My head is a wastebasket
(13) of worn ideas.

(14) Press my fingers
(15) and in my eyes appear
(16) credit and debit.
(17) Zing. Tinkle.
(18) My navel is a reject button.
(19) From my mouth issue canceled reams.
(20) Swollen, heavy, rectangular
(21) I am about to be delivered
(22) of a baby
(23) xerox machine.
(24) File me under W
(25) because I wonce
(26) was
(27) a woman.

1. This poem is an example of

    (1) rhymed and metered verse.

    (2) blank verse.

    (3) free verse.

    (4) visual poetry.

2. How is this poem organized? Is there logic to its structure?

3. List at least three similes or metaphors.

4. List the instances of onomatopoeia.

5. How does the speaker feel about being a secretary? How can you tell?

Answers

Poem A

1. The "old Lie" is that "It is sweet and fitting to die for one's country"—that dying in war is noble.

2. The person being addressed is (2), a civilian. It cannot be (1), a soldier, because the soldier would have experienced something similar to the speaker; nor can it be (4), the enemy, for the same reason. Between (2) and (3), choice (2), is a much better answer since civilians who are not fighting, who do not know the grim realities of war (whereas a reporter very well might), would be most likely to believe in the "old Lie."

3. The soldier dies from inhaling poison gas that came from the Five-Nines (the canisters that contained the gas).

4. In line 1, the men are "Bent double, like old beggars under sacks" (simile); in line 2, they are "coughing like hags" (simile); in line 12, the man who did not get his helmet on in time flounders "like a man in fire or lime" (simile); in line 14, the speaker with his helmet on during the gas attack sees "[a]s under a green sea" (simile); in line 19, the dying man's eyes are "writhing in his face" (personification); and in lines 20 and 23, the dying man's face is "like a devil's sick of sin" (simile), and his lungs are "[o]bscene as cancer" and "bitter as the cud / Of vile" (simile).

5. The rhyme scheme is *ababcdcd efef gh gh ijijklklmnmn*.

6. The metrical pattern is iambic pentameter.

7. The theme of the poem can be discerned from the closing stanza and the graphic description of the dying man. It is quite clear that the message of the poem is that it is *not* fitting and *not* sweet to die for one's country. Clearly, this soldier's death from gas is anything *but* sweet; it is utterly agonizing. The "old Lie" contends that dying in battle is something of a duty and an honor (see lines 25 and 26); yet, as seen from the perspective of one who watches a fellow soldier die, dying in battle is horror, not honor.

Poem B

1. The persona is speaking to Death.

2. Donne compares Death to (2), sleep, in line 5: "From rest and sleep, which but thy pictures be"; and again in lines 11 and 13: "And poppy, or charms can make us sleep as well" and "One short sleep past, we wake eternally."

3. Death is a slave "to fate, chance, kings, and desperate men" (line 9). This means that Death does not control when people die; rather, fate, chance, kings and desperate men decide when people will die. Instead of Death being in control, these things control Death.

4. In lines 13 and 14, "One short sleep past, we wake eternally, / And death shall be no more; Death, thou shalt die," the poet is suggesting that when we die, we actually move on to another life: the life of the soul, which, according to the Christian tradition, lives eternally with God. Thus, though our bodies may die, our souls actually live forever, where there is no more death, so death "shalt die," for it shall no longer exist. (Notice that death is no longer capitalized to show it has been reduced in importance.)

5. This poem is (3), an Italian sonnet.

Poem C

1. This poem is (3), free verse. There is no rhyme scheme or metrical pattern.

2. Although this poem is free verse, there is some logic to its structure. All of the lines but two—lines 5 and 19—are short, and many of the sentences are short, some even one word. This creates a choppy rhythm to the poem, a mechanical sound.

3. There are several metaphors: In line 1, "My hips are a desk"; in line 5, "My breasts are wells of mimeograph ink"; line 6 compares her feet to a chair, for they "bear casters"; lines 8–13 are all metaphors comparing the speaker's head to a file, switchboard and wastebasket; lines 14–16 compare her body to an adding machine; and finally in line 18, "My navel is a reject button."

4. In lines 7 and 17, "Buzz. Click," and "Zing. Tinkle," are instances of onomatopoeia.

5. The speaker seems to feel that by being a secretary she has *become* the machinery and supplies that she works with; that being a secretary has taken away her womanhood and made her a machine. Rather than giving birth to a child, she now gives birth to a xerox machine; her hips are the desk where she sits, her hair the rubber bands she uses to hold things together. This suggests that she works so hard that she has little or no time to be a mother or a woman.

REVIEW

Poetry is a form of writing that must be *heard* as well as read. Poems are usually brief, intense combinations of words and sentences that are often, but not always, in verse form. To understand poems we must look at elements of both sound and sense.

Poems, whether they rhyme or not, often have alliteration (repetition of sounds). The rhythm of a poem is determined by the meter (number of feet per line), which is often fixed according to a specific pattern, the rhyme, and the poet's diction.

The elements of poetry are essentially the same as the elements of fiction, with a few important differences. The *narrator* of fiction becomes a *persona* in poetry. In poems, because every word counts, we must pay particular attention to language and style. And because poems are unique in form, we must pay particular attention to structure as well.

Poems can be narrative, argumentative, imagistic, or visual; they can also be elegies or odes. They may be rhymed and metered, metered only (blank verse), or free verse (the rhyme and meter scheme, if any, is imposed by the author). Free-verse poems are usually organized by repetition or theme.

Interpreting Literature
and the Arts

Reading Drama

INTERPRETING LITERATURE AND THE ARTS

READING DRAMA

DEFINING DRAMA

Drama is one of the oldest ways of making sense of human experience. It is really the oldest form of storytelling. Before we had language, we were acting out our experiences and sharing our stories with each other.

Just as poetry is different from prose because of its form and emphasis on sound (it is meant to be read aloud), drama is different because it is literature that is meant to be performed. True drama (with the exception of "closet dramas") requires an audience. It is a social event in which the members of the audience are immediate witnesses to the action on the stage, and their energy is exchanged with the energy of the actors on stage before them.

Drama, which comes from the Greek word *dran*, meaning *to do* or *to act*, is composed of three essential parts:

- **dialogue** and/or **monologue**;
- **stage directions**; and
- an **audience**.

Because drama is meant to be performed, there is usually no narration—instead, the characters speak directly to each other in front of a live audience. Everything we learn we learn through their exchange of **dialogue** (or, in the case of only one character speaking to the audience, **monologue**). What the characters say to each other, then, becomes the main vehicle for meaning, for it is through language that characters represent how they see themselves and their world.

Look, for example, at the following dialogue from the opening scene in Sophocles' famous tragic play *Antigone*. Ismene and Antigone discuss their brothers, Eteocles and Polyneices, who have killed each other in a battle over control of the city of Thebes. With their deaths, Creon, their uncle, has become king. He has just forbidden the burial of Polyneices, because Polyneices had attacked Thebes. Their religion, however, made it a sin to leave someone unburied. Read the following dialogue to see how differently Antigone and Ismene feel about the situation:

ANTIGONE: Listen, Ismene:
  Creon buried our brother Eteocles
  With military honors, gave him a
    soldier's funeral,
  And it was right that he should; but

Polyneices,
Who fought as bravely and died as miserably,—
They say that Creon has sworn
No one shall bury him, no one mourn for him,
But his body must lie in the fields, a sweet treasure
For carrion birds to find as they search for food.
That is what they say, and our good Creon is coming here
To announce it publicly; and the penalty—
Stoning to death in the public square! There it is,
And now you can prove what you are:
A true sister, or a traitor to your family.

ISMENE: Antigone, you are mad! What could I possibly do?

ANTIGONE: You must decide whether you will help me or not.

ISMENE: I do not understand you. Help you in what?

ANTIGONE: Ismene, I am going to bury him. Will you come?

ISMENE: Bury him! You have just said the new law forbids it.

ANTIGONE: He is my brother. And he is your brother, too.

ISMENE: But think of the danger! Think what Creon will do!

ANTIGONE: Creon is not strong enough to stand in my way.

ISMENE: Ah, sister!
Oedipus died, everyone hating him
For what his own search brought to light, his eyes
Ripped out by his own hand; and Jocaste died,
His mother and wife at once: she twisted the cords
That strangled her life; and our two brothers died,
Each killed by the other's sword. And we are left:
But oh, Antigone,
Think how much more terrible than these
Our own death would be if we should go against Creon
And do what he has forbidden! We are only women,
We cannot fight with men, Antigone!
The law is strong, we must give in to the law
In this thing, and in worse. I beg the Dead
To forgive me, but I am helpless: I must yield
To those in authority. And I think it is dangerous business
To be always meddling.

ANTIGONE: If that is what you think
I should not want you, even if you asked to come,
You have made your choice, you can be what you want to be.
But I will bury him; and if I must die,
I say that this crime is holy: I shall lie down
With him in death, and I shall be as dear
To him as he to me.
   It is the dead,
Not the living, who make the longest demands:
We die for ever...
   You may do as you like
Since apparently the laws of the gods mean nothing to you.

ISMENE: They mean a great deal to me; but I have no strength
To break laws that were made for the public good.

ANTIGONE: That must be your excuse, I suppose. But as for me,
I will bury the brother I love.

[trans. Dudley Fitts and Robert Fitzgerald]

Questions

1. How does Ismene feel about burying Polyneices? Why?

2. How does Antigone feel about burying Polyneices? Why?

Answers

We can see from this dialogue that these two characters have two very different views of their worlds. For Ismene, it is more important to obey societal laws, laws "made for the public good," than to give her brother a proper burial according to the rules of their religion. Antigone, on the other hand, does not agree that they "must give in to the law" and "yield to those in authority." Instead, she will risk her life to obey the laws of the gods and honor her brother.

We can also see from this passage how differently these two characters feel about their position in society as women. Antigone says that Creon is "not strong enough" to stand in her way; Ismene, on the other hand, says "We are only women, / We cannot fight with men." Thus, Sophocles has used this opening dialogue to establish the beliefs of these two women, particularly the strong-willed Antigone, whose clash with Creon is the central action of the play.

In contrast to dialogue, a **monologue** is a play or part of a play in which one character speaks directly to the audience to reveal his or her thoughts. Look at the following excerpt from Jane Martin's *Rodeo* for example (1981). Big Eight—the character—is a former rodeo star who has been fired because she is not "modern" enough for the new owners of the rodeo.

> Used to be people come to a rodeo had a horse of their own back home. Farm people, ranch people—lord, they *knew* what they were lookin' at. Knew a good ride from a bad ride, knew hard from easy. You broke some bones or spent the day eatin' dirt, at least you got appreciated.
>
> Now they bought the rodeo. Them. Coca-Cola, Pepsi Cola, Marlboro damn cigarettes. You know the ones I mean. Them. Hire some New York faggot t' sit on some ol' stuffed horse in front of a sage brush photo n' smoke that junk. Hell, tobacco wasn't made to smoke, honey, it was made to chew. Lord wanted ya filled up with smoke he would've set ya on fire. Damn it gets me!
>
> There's some guy in a banker's suit runs the rodeo now. Got him a pinky ring and a digital watch, honey. Told us we oughta have a watchamacallit, choriographus or somethin', [...] Wants us to ride around dressed up like Mickey Mouse, Pluto, crap like that. Told me I had to haul my butt through the barrel race done up like Minnie damn Mouse in a tu-tu. Huh uh, honey!

Questions

1. How has the rodeo changed? What did it used to be like? How is it now?

2. What is it that "gets" Big Eight? What upsets her, and why?

Answers

1. The spectators at the rodeo used to be people who "*knew* what they were lookin' at" because they had horses of their own. They knew the skills required of the rodeo performers and could appreciate the rodeo for the talent of the riders, not the visual spectacle. "Now," though, the rodeo is commercialized, with sponsors like cola companies and with riders dressed up like Disney characters. It is also choreographed instead of natural.

2. Although Big Eight says "Damn it gets me!" after she complains about the cigarettes, what she is really upset about is the way the rodeo has changed. The cigarettes are just one symptom. Big Eight is upset because the rodeo has been taken over by "some guy in a banker's suit" who wants the rodeo to be a commercial production, not a natural test of a rider's skill. He wants everything to be staged, so to speak, and the riders to be cartoon characters instead of real people.

Language in drama—direct speech in dialogue and monologue—is the main way we are able to learn about the characters. Because there is usually no narrator in drama, there is no outsider to tell us what the characters really think and feel. They must tell us themselves through their words and actions.

Stage directions, the second essential component of drama, are the instructions provided by the playwright that tell the actors and directors how to perform the action and dialogue. Stage directions usually include specifications about costume, props, lighting, blocking (where the characters stand and how they should move on stage), and tone (how the characters should say certain lines). For example, a character may say a line "briskly" or "tenderly"; directions may tell a character to whisper or shout, to move closer or exit the stage.

The stage directions help convey the theme of the play in many ways, particularly through manipulation of tone and setting. The very first stage directions in a play usually describe the way the stage should be arranged and the appearance and mannerisms of the characters. The stage directions for *Rodeo*, for example, read:

> *A young woman in her late twenties sits working on a piece of tack. Beside her is a Lone Star beer in the can. As the lights come up we hear the last verse of a Tanya Tucker song or some other female country-western vocalist. She is wearing old worn jeans and boots plus a long-sleeved workshirt with the sleeves rolled up. She works until the song is over and then speaks.*

These stage directions tell the director what Big Eight should look like and what props (furniture and other items) should be with her on stage. But they also tell us much, much more. They reveal a good deal about Big Eight's character—what she likes, what she does, and what's important to her.

When there are very few stage directions, as is often the case in Shakespeare's plays and in the Greek tragedies, it is up to the director to determine where and how the characters move and speak, how they dress, and how to arrange the stage.

The **audience**, of course, is the final essential ingredient—for without an audience, a play cannot be produced. Plays are meant not only to indulge in relationships *on* the stage but also in relationships *between* the people on the stage and the members of the audience. There is constant feedback and exchange between the two.

So does this mean we can only see plays and not read them? Of course not. There is much to be gotten out of simply reading a play. In fact, when we read, we have the luxury—because of the lack of immediacy and the ability to re-read—to glean more meaning from the dialogue than we might had we seen it live. However, because we're not seeing the play acted out for us, we must do the work of the directors and actors ourselves, so to speak. We, as the readers, must bring the play to life by "staging" it inside our heads and creating a sort of **mental theater**, or theater of the mind. To do this, we must pay particular attention to the stage directions so we can see how things are supposed to happen on stage and how things sound.

ELEMENTS OF DRAMA

Again, drama is composed essentially of the same eight elements that make up fiction: **plot, character, setting, point of view, tone, language** and **style, symbolism,** and **theme**. But when we read drama, there are a few differences to take into consideration.

First, more so than in fiction, **action** is the driving force of the **plot**. "The essence of a play," said Aristotle, the first literary critic of the Western world, "is action." Of course, the action builds up to a **conflict**, which is at its peak in the **climax**, just as it is in fiction. However, in fiction, we can be "sidetracked" by the narrator—detoured from the action by description, detail, by getting inside characters' heads to hear their thoughts and learn about their backgrounds. In a play, even one with a narrator, there are no such "digressions": action must continue to occur and occur quickly, or else a live audience will certainly lose interest in the play.

Second, we need to consider **point of view**. Because there is no intermediary between story and reader, we don't have just one voice telling the story. In a play, even if there is a character who seems to serve as a narrator, like Tom in Tennessee William's play *The Glass Menagerie*, we still see the action represented for itself—we don't hear the story filtered through someone's point of view. Instead, we have what we call the **dramatic point of view**, and we are free to come to our conclusions based on the action and dialogue alone, not on any narrator's interpretation of the action. In other words, the dramatic point of view generally provides us with greater objectivity.

(We must keep in mind, however, that when characters tell us things about their past, for example, they can be telling us that information from their perspective—so all dialogue must be understood in the context of the characters who are speaking.)

Although there's no traditional narrator in drama, that doesn't mean we can't get inside the characters' heads. We learn about them through their dialogue and their actions, and when a playwright wants us to hear the private, inner thoughts of a character, he or she will use **soliloquy**, which allows the character to speak aloud on stage and reveal his or her thoughts as if no one is listening (essentially, the character is thinking aloud). In a **monologue**, the character is speaking directly to the audience, quite conscious of having a listener. In a soliloquy, however, the character speaks as if he or she is alone.

In *Hamlet*, for example, Hamlet's mother has married his uncle only a few months after his father's suspicious death. After his mother and stepfather try to cheer him up, he is left alone in his room and gives the following soliloquy expressing his anger at this hasty marriage:

> HAMLET: [...] That it should come to this!
> But two months dead! nay, not so much, not two:
> So excellent a king; that was, to this,
> Hyperion to a satyr: so loving to my mother,
> That he might not beteem the winds of heaven
> Visit her face too roughly. Heaven and earth!
> Must I remember? why, she would hang on him,
> As if increase of appetite had grown
> By what it fed on: and yet, within a month—
> Let me not think on't—Frailty, thy name is woman!—
> A little month, or ere those shoes were old
> With which she follow'd my poor father's body,

Like Niobe, all tears:—why she, even
   she,—
O God! a beast that wants discourse of
   reason
Would have mourn'd longer,—married
   with my uncle,
My father's brother, but no more like
   my father
Than I to Hercules: within a month;
Ere yet the salt of most unrighteous
   tears
Had left the flushing in her galled eyes,
She married. O, most wicked speed, to
   post
With such dexterity to incestuous
   sheets!
It is not, nor it cannot come to good:
But break, my heart, for I must hold my
   tongue!

These are thoughts Hamlet cannot express in front of others, especially his mother and stepfather, so he waits until he is alone to curse his mother's "frailty"—that she did not mourn his father long enough and married his brother with "most wicked speed."

An **aside** is a sort of blend between a soliloquy and a monologue. In an aside, a character shares his or her thoughts (usually about something that's happening on stage at the time) with the audience, but not with the other characters. The character reveals what he or she is thinking so we can know, but the other characters are not meant to hear.

For example, when Hamlet's new stepfather, Claudius, calls Hamlet "my cousin Hamlet, and my son," Hamlet makes the following aside: "A little more than kin, and less than kind." In other words, they've become too close—more than relatives with this double relationship of cousins (uncle/nephew) and father/son. And this relationship is "less than kind," because it is not normal, not good, and not considerate of others. This is especially true because we find out later that Claudius did indeed murder Hamlet's father—less than kind indeed.

In drama, **setting**—because it is part of the physical reality of the play—is very important. The setting depicted by the playwright reflects the playwright's philosophy about experience and understanding. For example, a realist like Henrik Ibsen, because he believed drama should be as true to life as possible, set his stage as realistically as possible. In addition, the action in his plays takes place in a limited time period so that it is almost as if we are spying on the characters—as if we can see through the wall of the room in which the action takes place.

Many dramatists make frequent use of **dramatic irony** as the controlling tone. It is employed often, perhaps, because it is easy for members of the audience to learn things about some characters that other characters do not know; thus, we understand things that other characters do not.

For example, in Act I of *A Doll House*, by Henrik Ibsen, we learn Nora's secret: that several years ago she forged her father's signature on a loan document to borrow money that she needed to help her husband, Helmer, recover from a serious illness. She borrowed that money from Krogstad, who has been fired by Helmer now that Helmer has been promoted to manager at the bank where they work. Krogstad has decided to blackmail Nora in an effort to keep his job at the bank. In order for Nora to keep her secret—which she must do, since Helmer loathes the idea of borrowing money—she must convince Helmer to take Krogstad back. In this scene, Helmer explains to Nora why this is impossible: because Krogstad has a seedy past.

[Trans. Rolfe Fjelde]

NORA: But tell me, was it really such a crime that this Krogstad committed?

HELMER: Forgery. Do you have any idea what that means?

NORA: Couldn't he have done it out of need?

HELMER: Yes, or thoughtlessness, like so many others. I'm not so heartless that I'd condemn a man categorically for just one mistake.

NORA: No, of course not, Torvald!

HELMER: Plenty of men have redeemed themselves by openly confessing their crimes and taking their punishment.

NORA: Punishment—?

HELMER: But now Krogstad didn't go that way. He got himself out by sharp practices, and that's the real cause of his moral breakdown.

NORA: Do you really think that would–?

HELMER: Just imagine how a man with that sort of guilt in him has to lie and cheat and deceive on all sides, has to wear a mask even with the nearest and dearest he has, even with his own wife and children. And with the children, Nora—that's where it's most horrible.

NORA: Why?

HELMER: Because that kind of atmosphere of lies infects the whole life of a home. Every breath the children take in is filled with the germs of something degenerate.

NORA: [*coming closer behind him*] Are you sure of that?

HELMER: Oh, I've seen it often enough as a lawyer. Almost everyone who goes bad early in life has a mother who's a chronic liar.

NORA: Why just—the mother?

HELMER: It's usually the mother's influence that's dominant, but the father's works in the same way, of course. Every lawyer is quite familiar with it. And till this Krogstad's been going home year in, year out, poisoning his own children with lies and pretense; that's why I call him morally lost. [*Reaching his hands out toward her.*] So my sweet little Nora must promise me never to plead his cause. Your hand on it. Come, come, what's this? Give me your hand. There, now. All settled. I can tell you it'd be impossible for me to work alongside of him. I literally feel physically revolted when I'm anywhere near such a person.

Did you notice the irony in this passage? Did you realize that what Helmer said about Krogstad applies, without him even realizing it, to his wife? *He* doesn't know that Nora has committed the same crime as Krogstad, and he doesn't realize that she has likewise "infected" their own household with lies. That is why she asks the kind of questions she does in this passage.

The greatest irony in this passage is the last line. Helmer says he is "physically revolted" when he's near a person like Krogstad. Yet, he is holding Nora's hands as he says this. *We* know the significance and feel the impact of these words, as does Nora, but he does not.

A BRIEF HISTORY OF DRAMA

Drama, as we have said, is one of the oldest ways of making sense of our experiences—essentially the first form of storytelling. Early cultures developed dramatic storytelling rites in which "actors" would describe hunts, tribal wars, and other conflicts. As cultures and societies evolved, dramas became concerned with cultural heroes and gods, as well as celebrations of stories, people and events. Around the fifth century B.C., however, Greek drama evolved—and this is really where drama as we know it began.

Perhaps the most renowned of the ancient Greek dramatists is Sophocles, who wrote the Oedipus cycle (*Oedipus Rex, Oedipus at Colonus, Antigone*). Read the following passage from *Antigone* and see what you can learn about Greek culture from the dialogue:

> CREON: (*slowly, dangerously*) And you, Antigone,
> you with your head hanging,—do you confess this thing?
> ANTIGONE: I do. I deny nothing.
> CREON: (*to Sentry*): You may go.
> *Exit* Sentry.
> (*To Antigone*) Tell me, tell me briefly:
> Had you heard my proclamation touching this matter?
> ANTIGONE: It was public. Could I help hearing it?
> CREON: And yet you dared defy the law.
> ANTIGONE: I dared.
> It was not God's proclamation. That final Justice
> That rules the world below makes no such laws.
> Your edict, King, was strong,
> But all your strength is weakness itself against
> The immortal unrecorded laws of God.
> They are not merely now: they were, and shall be,
> Operative for ever, beyond man utterly.
> I knew I must die, even without your decree:
> I am only mortal. And if I must die
> Now, before it is my time to die,
> Surely this is no hardship: can anyone
> Living, as I live, with evil all about me,
> Think Death less than a friend? This death of mine
> Is of no importance; but if I had left my brother
> Lying in death unburied, I should have suffered.
> Now I do not.
> You smile at me. Ah Creon,
> Think me a fool, if you like; but it may well be
> That a fool convicts me of folly.
> CHORUS: Like father, like daughter: both headstrong, deaf to reason!
> She has never learned to yield.
> CREON: She has much to learn.
> The inflexible heart breaks first, the toughest iron
> Cracks first, and the wildest horses bend their necks
> At the pull of the smallest curb.
> Pride? In a slave?
> This girl is guilty of double insolence,
> Breaking the given laws and boasting of it.
> Who is the man here,
> She or I, if this crime goes unpunished?

Questions

1. What is the conflict between Creon and Antigone?

2. Why is Antigone not afraid to die?

3. What does Creon seem to be concerned with in the last two lines?

4. Do you think Antigone is a typical Greek woman? Why or why not?

Answers

1. The conflict between Creon and Antigone is similar to the one we saw earlier between Ismene and Antigone. Creon wants Antigone to obey his laws, but she has already broken his law forbidding the burial of her brother, Polyneices. She believes that she must obey the laws of the gods rather than the laws of man. Creon cannot accept this; he is outraged not only that she dared to break his law but that she does not repent.

2. Antigone is not afraid to die because she has obeyed the laws of God.

3. Creon seems to be concerned with his manhood. He is afraid that if he does not punish Antigone, he will be less of a man.

4. Antigone is not a typical Greek woman. In Greek society, women were considered third-class citizens, even if they were women in a noble household. Thus, a more typical Greek woman would be Ismene, whom we saw earlier. Greek women typically would not dare to challenge men, especially men of power like Creon.

The Greeks, who made drama the art form that it is today, held their plays in huge amphitheaters that could hold crowds of up to 14,000 people (compare that to the average theater of today, which usually holds no more than 300). Because they were limited in terms of stage props and machinery, one element of Greek drama is **unity**: unity of time, of place, and of action. The closely linked events acted out on stage always took place within a limited period of time (24 hours, often) and in a single setting (a palace, for example). Another element of Greek drama was the chorus—the singing actors who offered the audience a commentary on the action.

One of the greatest achievements of Greek theater was the perfection of the **tragic play**. A **tragedy**—literally, a play or event that causes great sadness—is a play that presents a noble character's fall from greatness. All of the characters in Greek plays are men and women of nobility—kings, queens, their families and attendants. In the course of the play, the protagonist of a tragedy—a king, for example—tumbles from his or her high position as a character of great respectability to a low position as a character to be scorned. This fall usually takes place for one of two reasons (or a combination of the two): the character's **tragic flaw** and/or **fate**.

The tragic flaw is the characteristic that is often part of what makes the character so great but also what drives the character to make bad or wrong decisions. Pride is such a characteristic. Absolutism is Creon's tragic flaw in *Antigone*, which makes the passage quoted earlier quite ironic, since he accuses Antigone of the same flaw. In *Antigone*, Creon was a praised as a wise ruler because he put the welfare of the state before the welfare of any individual. He then refused to make an exception when Antigone, his niece and son's fiancee, broke the law. As a result Antigone, Creon's son, and Creon's wife all killed themselves by the end of the play. Only Creon was left to survey the destruction he caused.

Still, tragedy, despite its heaviness, carries a positive message. A tragedy shows us what it means to suffer, what it means to be human—to make a mistake and suffer the dire consequences. In addition, in a true tragedy, the tragic character always accepts responsibility for what he or she has done. Even though it's too late to take back the tragic events that have occurred, at least we know that the tragic character has learned from his or her mistakes, and there is hope that he or she will be a wiser, more just person in the future. Creon, for example, has learned (albeit too late) to be more flexible. Hence, we can call this person the **tragic hero**.

This is why, in Arthur Miller's *Death of a Salesman*, Willy Loman is *not* a tragic hero. Instead, he is a **pathetic** character. He refuses, up to his suicide, to acknowledge his errors and his failures. His son Biff, on the other hand, *accepts* responsibility for what he has done to keep Willy's dreams alive. Also, Biff attempts, throughout the play, to get Willy to accept responsibility. Thus, Biff is a true tragic hero.

At the opposite end of the spectrum, of course, is the **comedy**. A comedy, as a rule, has a happy ending. Rather than ending in death, sadness, and separation, a comedy ends in union and happiness. The "jokes" in comedies often arise from **puns** (plays on words) and double meanings. They also come from overturning our expectations. Take Woody Allen's one-act play "Death Knocks," for example. In this play, Death knocks on Nat Ackerman's door—or rather, climbs in his window. Normally, we'd expect Death—the Grim Reaper—to be a sombre, frightening, powerful character of few words other than commands. But, as you can see from the opening stage directions and the first lines of dialogue from the play, this is not at all the case:

[*Climbing awkwardly through the window is a sombre, caped figure. The intruder wears a black hood and skintight black clothes. The hood covers his head but not his face, which is middle-aged and stark white. He is something like NAT in appearance. He huffs audibly and then trips over the windowsill and falls into the room.*]

DEATH: [*for it is no one else*] Jesus Christ, I nearly broke my neck.
NAT: [*watching with bewilderment*] Who are you?
DEATH: Death.
NAT: Who?
DEATH: Death. Listen—can I sit down? I nearly broke my neck. I'm shaking like a leaf.
NAT: Who *are* you?
DEATH: *Death.* You got a glass of water?
NAT: Death? What do you mean, Death?
DEATH: What is wrong with you? You see the black costume and the whitened face?
NAT: Yeah.
DEATH: Is it Halloween?
NAT: No.
DEATH: Then I'm Death. Now can I get a glass of water—or a Fresca?
NAT: If this is some joke—
DEATH: What kind of joke? You're fifty-seven? Nat Ackerman? One eighteen Pacific Street? Unless I blew it—where's that call sheet? [*He fumbles through his pocket, finally producing a card with an address on it. It seems to check.*]
NAT: What do you want with me?
DEATH: What do I want? What do you think I want?
NAT: You must be kidding. I'm in perfect health.
DEATH: [*unimpressed*] Uh-huh. [*Looking around.*] This is a nice place. You do it yourself?
NAT: We had a decorator, but we worked with her.
DEATH: [*looking at picture on the wall*] I love those kids with the big eyes.
NAT: I don't want to go yet.
DEATH: *You* don't want to go? Please don't start in. As it is, I'm nauseous from the climb.
NAT: What climb?
DEATH: I climbed up the drainpipe. I was trying to make a dramatic entrance. I see the big windows and you're awake reading. I figure it's worth a shot. I'll climb up and enter with a little—you know…[*Snaps fingers.*] Meanwhile, I get my heel caught on some vines, the drainpipe breaks, and I'm hanging by a thread. Then my cape begins to tear. Look, let's just go. It's been a rough night.

Notice how Death here is portrayed not as a frightening, ominous character but rather a clumsy, casual, hassled character who's even unsure if he's at the right address. Allen has completely challenged our idea of what a visit by the Grim Reaper would be like, and in doing so, makes the notion of dying a little less frightening.

A **melodrama** is a "tragedy" that has a happy ending, thereby ruining the effect of a true tragedy. Many plays are **tragicomedies**, which differ from melodrama. The ending is still tragic, but interspersed in the play are some comic scenes to ease the intensity of emotion a tragedy arouses.

After Greek drama, the next step in the evolution of Western theater was the medieval **mystery** and **morality plays**. These were the dramatic versions of fables and parables. The characters in these plays were allegorical or symbolic, and the plays were highly didactic to teach the audience members religious and moral ideas.

During the Renaissance, patronism—the sponsorship of artists by members of the nobility—largely controlled the direction of drama. Playwrights were hired by kings and queens, by dukes and duchesses, to write plays that often glorified them and their country. That's why there were so many history plays written during this time. It was at this time, of course, that Shakespeare, the dramatic genius, wrote his plays and poems.

It was only in the late seventeenth century that theaters became enclosed and the stage as we know it was developed. Our current "picture frame" stage is like a room where the fourth wall has been removed, so that we are almost like voyeurs peeking into someone's private life.

At the turn of the nineteenth century, one of the most important movements in drama, **realism**, developed. In reaction to the unrealistic Romantic plays of the eighteenth and nineteenth centuries, playwrights wanted to represent real life as accurately as possible—with real, "ordinary" characters dealing with the ordinary struggles of life, like birth, death, love, poverty, and unhappy marriages. No longer did plays focus on the "extraordinary" events like wars and political intrigues that characterized the plays of Shakespeare's time. The object was to open the audience's eyes to the truth of their lives. In addition, realists began to use "real" dialogue—that is, their characters spoke naturally, as real people do, rather than in verse, as Shakespeare's characters do. One of the greatest playwrights of this era was Ibsen, who popularized the social drama or problem play.

Modern (post-realist) theater rejects the idea that a play can or should mimic life. These plays often emphasize the fact that they are, indeed, plays, which are artificial representations of reality. They draw attention to their conventions and devices (such as props) and comment on reality rather than trying to imitate it. Many modern playwrights challenge the notion that reality is ordered and meaningful (and therefore imitatable), and from this idea emerges the **antihero**—a main character who is more pathetic than admirable, more destructive than constructive. Willy Loman is a perfect example.

Some of the plot structure of *Death of a Salesman* is unrealistic. For example, the plot is layered so that past and present are constantly overlapping. In fact, on occasion, Willy Loman is both in the present and the past. Now, this might not seem "real" to you, but if you think about it, we function in many "tenses" at the same time. Even as we talk to someone now, our minds may be lost in a past memory or moment or in a dream about the future. So, though we do not call this "realism," it is one very effective way of reflecting and interpreting reality.

Today's playwrights often return to the unity and simplicity of Greek plays or the realism of Ibsen's era. Others, however, continue to experiment with theater to find new and exciting ways to express their ideas on stage.

Practice: Applying Comprehension Strategies

DIRECTIONS: Read the following passages. Apply the four reading strategies you've learned as well as your vocabulary skills. Write in the margins and mark up the text as you go. Then answer the questions following each passage.

PASSAGE A

The following passage is an excerpt from Susan Glaspell's one act play, "Trifles" (1916). Note: Minnie Foster is Mrs. Wright's maiden name.

MRS. HALE: She liked the bird. She was going to bury it in that pretty box.

MRS. PETERS: [*In a whisper*] When I was a girl—my kitten—there was a boy took a hatchet, and before my eyes—and before I could get there—[*Covers her face an instant.*] If they hadn't held me back I would have—[*Catches herself, looks upstairs where steps are heard, falters weakly*]—hurt him.

MRS. HALE: [*With a slow look around her*] I wonder how it would seem never to have had any children around. [*Pause.*] No, Wright wouldn't like the bird—a thing that sang. She used to sing. He killed that, too.

MRS. PETERS: [*Moving uneasily*] We don't know who killed the bird.

MRS. HALE: I knew John Wright.

MRS. PETERS: It was an awful thing was done in this house that night, Mrs. Hale. Killing a man while he slept, slipping a rope around his neck that choked the life out of him.

MRS. HALE: His neck. Choked the life out of him.

[*Her hand goes out and rests on the bird-cage.*]

MRS. PETERS: [*With rising voice*] We don't know who killed him. We don't *know*.

MRS. HALE: [*Her own feeling not interrupted*] If there'd been years and years of nothing, then a bird to sing to you, it would be awful—still, after the bird was still.

MRS. PETERS: [*Something within her speaking*] I know what stillness is. When we homesteaded in Dakota, and my first baby died—after he was two years old, and me with no other then—

MRS. HALE: [*Moving*] How soon do you suppose they'll be through, looking for the evidence?

MRS. PETERS: I know what stillness is. [*Pulling herself back.*] The law has got to punish crime, Mrs. Hale.

MRS. HALE: [*Not as if answering that*] I wish you'd seen Minnie Foster when she wore a white dress with blue ribbons and stood up there in the choir and sang. [*A look around the room.*] Oh, I *wish* I'd come over here once in a while! That was a crime! That was a crime! Who's going to punish that?

1. Who killed Mr. Wright? How?

2. Why did the murderer do it?

3. What was Minnie Foster like? How did she change?

4. What is Mrs. Peters' conflict?

5. Of what crime is Mrs. Hale guilty?

PASSAGE B

The following passage is an excerpt from the Greek playwright Aristophanes' *Lysistrata*.

COMMISSIONER: Very well.
   My first question is this: Why, so help you God,
   did you bar the gates of the Akropolis?
LYSISTRATA: Why?
   To keep the money, of course. No money, no war.
COMMISSIONER: You think that money's the cause of war?
LYSISTRATA: I do.
   Money brought about that Peisandros business
   and all the other attacks on the State. Well and good!
   They'll not get another cent here!
COMMISSIONER: And what will you do?
LYSISTRATA: What a question! From now on, we intend
   to control the Treasury.
COMMISSIONER: Control the Treasury!
LYSISTRATA: Why not? Does that seem strange? After all,
   we control our household budgets.
COMMISSIONER: But that's different!
LYSISTRATA: "Different?" What do you mean?
COMMISSIONER: I mean simply this:
   it's the Treasury that pays for National Defense.
LYSISTRATA: Unnecessary. We propose to abolish war.
COMMISSIONER: Good God.—And National Security?
LYSISTRATA: Leave that to us.
COMMISSIONER: You?
LYSISTRATA: Us.
COMMISSIONER: We're done for, then!
LYSISTRATA: Never mind.
   We women will save you in spite of yourselves.
COMMISSIONER: What nonsense!
LYSISTRATA: If you like. But you must accept it, like it or not.
COMMISSIONER: Why, this is downright subversion!
LYSISTRATA: Maybe it is.
   But we're going to save you, Judge.
COMMISSIONER: I don't *want* to be saved.
LYSISTRATA: Tut. The death-wish. All the more reason.
COMMISSIONER: But the idea of women bothering themselves about peace and war!
LYSISTRATA: Will you listen to me?
COMMISSIONER: Yes. But be brief, or I'll—
LYSISTRATA: This is no time for stupid threats.
COMMISSIONER: By the gods,
   I can't stand any more!
AN OLD WOMAN: Can't stand? Well, well.
COMMISSIONER: That's enough out of you, you old buzzard!
   Now, Lysistrata, tell me what you're thinking.
LYSISTRATA: Glad to.
   Ever since this war began
   We women have been watching you men, agreeing with you,
   keeping our thoughts to ourselves. That doesn't mean
   we were happy: we weren't, for we saw how things were going;
   but we'd listen to you at dinner
   arguing this way and that.
     —Oh you, and your big
   Top Secrets!—
   And then we'd grin like little patriots
   (though goodness knows we didn't feel like grinning) and ask you:
   "Dear, did the Armistice come up in

Assembly today?"
And you'd say, "None of your business! Pipe down!" you'd say.
And so we would.
AN OLD WOMAN: *I* wouldn't have, by God!
COMMISSIONER: You'd have taken a beating, then!
—Go on.
LYSISTRATA: Well, we'd be quiet. But then, you know, all at once
you men would think up something worse than ever.
Even *I* could see it was fatal. And, "Darling," I'd say,
"have you gone completely mad?" And my husband would look at me
and say, "Wife, you've got your weaving to attend to.
Mind your tongue, if you don't want a slap. 'War's a man's affair!'"
COMMISSIONER: Good words, and well pronounced.
LYSISTRATA: You're a fool if you think so. It was hard enough
to put up with all this banquet-hall strategy.
But then we'd hear you out in the public square:
"Nobody left for the draft-quota here in Athens?"
you'd say; and, "No," someone else would say, "not a man!"
And so we women decided to rescue Greece.
You might as well listen to us now: you'll have to, later.

1. Why does Lysistrata want to take control of the Treasury?

2. What does the Commissioner think of her plan?

3. Why was "not a man" left for the draft-quota?

4. Lysistrata and the women are wives of
   (1) farmers.
   (2) politicians.
   (3) teachers.
   (4) bankers.

5. Why didn't the wives say anything to the men?

6. How does Lysistrata plan to "rescue Greece"?

Answers

Passage A

1. Mrs. Wright (Minnie Foster) killed Mr. Wright by slipping a rope around his neck while he slept. She "choked the life out of him" like he choked the life out of her bird.

2. At first it seems that Mrs. Wright killed her husband because he killed her bird, which she loved and which was very much like her (she used to sing, too). But her anger about the murder of the bird is apparently a symptom of a greater problem: Mrs. Wright was lonely. We can see from the passage that the Wrights never had any children, so the pet meant a great deal to her. But Mr. Wright, the passage reveals, was a man who didn't like singing—either the bird's or his wife's. Thus in the Wright household there was "years and years of nothing," and then the bird. So when Mr. Wright killed the bird, it was "awful—still." That stillness, that loneliness, drove Mrs. Wright to kill the man who had made her feel so alone.

3. Mrs. Wright, when she was Minnie Foster, used to sing in the choir. This suggests that

she was a happy, social person. Since she married Mr. Wright, however, she stopped singing ("He killed that, too"), and, it is implied, stopped socializing.

4. Mrs. Peters has a very complex conflict. She is torn between understanding and condoning Mrs. Wright and upholding the law. She understands both how Mrs. Wright felt when her bird was killed (she remembers when the boy killed her kitten) and how Mrs. Wright felt when it was still ("I know what stillness is," she says). But she also wants to obey the law and punish whoever murdered Mr. Wright, for whatever reason. "The law has got to punish crime," she says.

5. Mrs. Hale is guilty of not coming to visit Mrs. Wright. She has not been a good neighbor; she has not kept the lonely Mrs. Wright company.

Passage B

1. Lysistrata wants to take control of the Treasury to stop the war.

2. The Commissioner thinks her plan is ridiculous ("the idea of women bothering themselves about peace and war!" he says) and that it will lead to the ruin of Greece ("We're done for, then!" he cries).

3. There was "not a man" left for the draft-quota because all the men had already been drafted and sent to war. That they need more men indicates that most of the men were still serving as soldiers or, more likely, they had already been killed.

4. Lysistrata and the other women overtaking the Treasury are the wives of (2), politicians. This becomes clear when Lysistrata explains to the Commissioner why they've taken over the Treasury. Their husbands talk about "Top Secrets" and the wives ask them, "did the Armistice come up in Assembly today?" It is clear that their husbands are senators and other government officials.

5. The wives didn't say anything to the men because they were expected not to interfere. When they asked about the Armistice, the men replied "None of your business! Pipe down!" It was clear that women were not expected to have opinions or interest in politics, and the women's place was to obey their husbands. When Lysistrata tried to offer her opinion about the war, her husband replied, "Mind your tongue, if you don't want a slap. 'War's a man's affair!'" The Commissioner agrees, so it is clear that women were expected to stay out of politics.

6. Lysistrata plans to "rescue Greece" by controlling the Treasury (because she believes money causes war) and thereby abolishing war.

REVIEW

Drama is literature that is meant to be performed. It requires dialogue and/or monologue, stage directions, and an audience. When we read drama, we should picture the stage and action in our heads.

The elements of drama are essentially the same as those of fiction: plot, character, setting, point of view, tone, language and style, symbolism, and theme. However, in drama, the point of view is very different. In drama, there is usually no narrator to serve as an intermediary between the story and the reader. With no "filter," we have a more objective view of the characters and their actions.

Dramatic irony is frequently employed by playwrights to control tone. Dramatic irony is

when we, as members of the audience, know something about the characters or action that other characters don't, such as Nora's crime.

Drama as we know it was developed by the Greeks, whose plays had unity of time, place, and action. The Greeks often wrote tragedies, plays that documented a noble character's downfall due to a tragic flaw, such as pride, and/or fate. Tragedies end in destruction, but also in hope, because the tragic hero accepts responsibility for what he or she has done.

Realism was an important movement in the history of drama. Whereas most plays up to the nineteenth century dealt with the extraordinary lives and conflicts of the nobility, realists wrote plays that dealt with real, "ordinary" people and their everyday lives and concerns. Playwrights strove to make their plays as believable and true to life as possible.

Since realism, playwrights have been experimenting with theater. Some plays call attention to the conventions of drama to emphasize that they are not attempting to accurately represent reality.

Interpreting Literature and the Arts

Reading Commentary

INTERPRETING LITERATURE AND THE ARTS

READING COMMENTARY

WHAT IS COMMENTARY?

Commentary, unlike prose, poetry, or drama, is not written for its own sake. Instead, its purpose is to illuminate or explain *other* works of literature and art.

Commentary literally *comments* on a work of art by reviewing and analyzing it for the reading audience. A commentator's job is to help us understand this work of art and "rate" its success and value. We read commentary, for example, to find out what a new book is about and how the reviewer reacts to it; to decide whether we should see movie A or movie B (we might ask, "Which got a better review?"). Commentary, then, serves as a "litmus test" for art.

Although there is commentary on all types of art—performance art (dance, music, theater), visual art (painting, sculpture, photography), and literature—we will focus here on book, movie, theater, music, and television reviews.

When reading commentary, one of the most important skills to master is to be able to distinguish clearly between **facts** and **opinions**. Commentary, by nature, is highly **subjective,** since the writer is sharing his or her personal response to the work. A good commentator, however, will always "defend" his or her response by clearly explaining *why*. He didn't like the movie, for example, because the script was clichéd, the characters were flat and the scenes were poorly directed. She loved the book because the characters were vivid and believable, the plot was captivating, and the language and style were original. We must remember, however, that these are all opinions, no matter how well the author may defend them. The **facts** are the elements that are not debatable. That so-and-so stars in the movie, for example, is a fact. That a movie is 135 minutes long and was directed by XYZ are also facts. That X, Y, and Z happen in the movie are facts as well. And these facts enable us to form our own opinions.

Another skill to master when reading commentary is noticing the commentator's use of language. Even in sentences where the author is not directly expressing an opinion, he or she will use words that express his or her attitude about the work of art being reviewed. For example, look at the following sentences. They have the same meaning but convey a different attitude:

a. This novel is 100 pages.
b. This masterpiece is 100 pages.
c. This failure is 100 pages.

The change of "novel" to "masterpiece" or "failure" is a conscious choice on the author's part to convey his or her feelings about those hundred pages. The commentator's diction, then, is very important.

BOOK REVIEWS

A book review usually begins with some summary of the text's plot and its major conflicts, as well as some background information on the author. Reviews comment on the quality and originality of the writing (the author's language and style) as well as the quality and originality of the story line or subject matter. Generally, reviews will take into consideration all of the elements of fiction and non-fiction that apply and highlight both what the author does well and what the author does not do so well. We must remember, of course, that just because a **reviewer** (also called **critic**) does not like a book doesn't mean that *we* will not enjoy it. Usually reviewers will point out both a text's strengths and its flaws and allow us to make this judgment for ourselves.

Read the following excerpt from Rand Richard Cooper's review of Geoffrey Becker's short story collection *Dangerous Men*:

Geoffrey Becker's first collection of stories appears under a misleading title. Its heroes—vulnerable boys, tender adolescents and twenty-something musicians—are hardly men at all, and none too dangerous either: a harmonica player returning from Europe with his girlfriend to an American woefully short on romance; two brothers running a small-time recording studio and vainly chasing the same *femme fatale*; a 15-year-old hitting the road with his hard-luck bluesman father. Readers expecting fast-paced action will be disappointed, for "Dangerous Men" focuses less on what happens than on what doesn't. These are stories about what is missed, lost, decided against or regretted, once the action is over.

Mr. Becker's book contains some typical first-time-out errors: anomalous tense shifts; leaps into sentimental lyricism; escape-hatch moves where the author simply tells us, in a kind of slap-dash notation, what a character is feeling. There's also a strong flavor of the cinematic, with short scenes everywhere. After a while, you start longing for looping flights of thought as well as the increasingly familiar moment-to-moment take on things.

Still, the best of these stories extend attractive invitations to our sympathies. Mr. Becker's portrayals of teen-agers are particularly deft. In "Magister Ludi," a 17-year-old girl named Duney minds the fort while her 15-year-old brother stages a rowdy party in the basement; here Mr. Becker perfectly pegs the conflicting impulses of condescension, envy, and disbelief with which an older sibling greets the emergence of a younger. "Taxes" takes us to a bleak corner of Brooklyn where a black kid named Pretzel, who acts as errand boy for an elderly Jewish tax preparer, is being pressured by an older brother to rip his employer off. Struggling to gauge loyalties and fix a future that might involve something other than crime, Pretzel is trying not to be bad in a place where it's hard to be good, and Mr. Becker paints that attempt with tender care.

Questions

1. Is the title of Mr. Becker's collection appropriate? Why or why not?

2. What kind of readers will not like the collection? Why?

3. What age are most of the main characters in the stories?

    (1) infants

    (2) 15–20

    (3) 30–50

    (4) over 60

4. What faults does Cooper find with Becker's collection? List them below:

    _____

    _____

    _____

5. "Deft" in the third paragraph means

    (1) sloppy.

    (2) quick.

    (3) happy.

    (4) skillful.

Answers

1. According to the author of the review, the title Mr. Becker has given to his collection is not appropriate ("misleading") because most of the characters in his stories are not men but teenagers and young adults, and his characters are "none too dangerous."

2. The author thinks that readers who like stories packed with action and adventure will be disappointed by Mr. Becker's collection. Becker's stories, he says, are more about what *doesn't* happen than what does.

3. Most of the main characters are (2), between 15–20. We can tell from the first paragraph of the review, which calls the characters "vulnerable boys, tender adolescents, and twenty-something musicians." Therefore, (1), (3) and (4) cannot be correct.

4. Cooper finds a few faults with Becker's collection: shifts in verb tense, sentimentalism, telling the reader "in a kind of slap-dash notation" how a character feels, and the overly-cinematic feel of the stories.

5. "Deft" means (4), skillful. Cooper calls Becker's "portrayals of teen-agers... particularly deft," and in the rest of the paragraph he describes how Becker "perfectly pegs the conflicting impulses" of one of his characters and how Becker "paints" a character's struggle "with tender care." The paragraph is entirely complimentary about Becker's ability to portray characters, so (1) and (2) are not correct answers, and (3) is not appropriate either.

THEATER REVIEWS

Theater reviewers take into consideration the elements of drama. Some questions a theater reviewer might answer are: Does the play engage the audience? Is it well-acted? Effectively directed? If it is a production of an old play, does it offer a new interpretation of the action or characters? How is it different from other productions? If it is a new play, what issues does the playwright address? Has he or she put together a convincing drama—are the characters real? How do we feel when we walk out of the theater? Is it a play I'll forget in minutes, or one that will haunt me for years to come?

The following review of the new play "Racing Demon" was written by Margo Jefferson. Read it carefully and answer the questions that follow.

You will be comfortable watching David Hare's "Racing Demon"; comfortably alert, comfortably stimulated—intellectually and dramatically; comfortably concerned with an Important Issue of the Day: the role of the church and of Christianity in society; the ruthless political maneuvers and machinations and the battles for power and domination that go on inside the church.

Of course, it's the Church of England. "Racing Demon," set in Britain, is one of three plays that David Hare has written that analyze, criticize and dramatize social and political institutions. [....]

There is suspense. What happens when a principled clergyman (played by Josef Sommer), a man of conscience but also a man who is depressed and spiritually depleted, comes up against a passionate and ruthless young evangelical (Michael Cumpsty) who believes that he and he alone can snare and save souls in a struggling and impoverished South London neighborhood?

What happens when the older clergyman is a member of the church's upper-middle-class-old-boy network, while the young evangelical is a working-class upstart? What will happen to the long-neglected wife of one (Kathleen Chalfant) and the recently cast-off lover of the other (Kathryn Meisle)?

Mr. Hare does not miss an issue: homosexuality, wife-beating, abortion and the ordination of women all fuel the plot. As well they should, though the latter two subjects are tossed into the exposition pretty hastily. And there is a lot of exposition here. Mr. Hare conducts strong, even gripping arguments, but he is like a journalist who has done so much solid reporting and research that he cannot bear to edit himself. He depends on brief soliloquies in the form of prayers to give us a glimpse of each man's inner life. He gives the women poetic declarations, usually at the end of a scene.

Questions

1. The audience watching "Racing Demon" will feel _____ watching Hare's play.

    (1) disturbed

    (2) at ease

    (3) light-hearted

    (4) upset

2. The main conflict in the play is between what two characters?

3. How do the characters reveal themselves?

4. What does Jefferson think Hare does well in the script?

   (1) Good arguments

   (2) Good dialect

   (3) Good figurative language

   (4) Good jokes

5. What does Hare not do that Jefferson thinks he should?

   (1) Joke more

   (2) Revise more

   (3) Edit more

   (4) Include more issues

6. Based on this commentary you can conclude that Jefferson thinks

   (1) a good play will not disturb you too much.

   (2) a good play will make you laugh a lot.

   (3) the theater should be comfortable.

   (4) a good play will disturb you.

Answers

1. The audience will feel (2), at ease. This comes across quite clearly in the opening sentence of Jefferson's review: "You will be *comfortable* watching David Hare's "Racing Demon," she says, and she repeats "comfortably" several times. "At ease" is the best synonym for "comfortable."

2. The main conflict in the play is between an older clergyman and a "passionate and ruthless" evangelist.

3. The characters in Hare's play reveal their inner feelings mostly through their prayers ("brief soliloquies in the form of prayers").

4. Jefferson thinks Hare has (1), good arguments, in his script. ("Hare conducts strong, even gripping arguments," she writes.)

5. Jefferson thinks that Hare should (3), edit more. She says he "is like a journalist who has done so much solid reporting and research that he cannot bear to edit himself." He has included too much, too many issues.

6. Based on this passage we can infer that Jefferson thinks (4), a good play will disturb you. We can infer this based largely on the repetition of "comfortably" in the first paragraph and the capitalization of "Important Issue of the Day," which seems to lightly mock Hare's attempt to address important issues without disturbing his audience too much.

MUSIC REVIEWS

Like all critics, music reviewers will consider both the work of art itself and its effect on the audience. Music critics might ask: Are the music and lyrics original? (Or, if it is a new production of old music, is it an original interpretation?) Are the musicians barely competent, or masters of their instruments? Does the vocalist yell, or croon? What is the mood of the music? Is it angry, happy, or hypnotic? What kind of presence do the musicians have on stage? And so on.

Take a look at this excerpt from Jon Pareles' review of a recent performance by the legendary Bruce Springsteen:

> For just over two hours, Mr. Springsteen stood alone with his guitar, singing

about people crushed by hard times and bad decisions. He prefaced many songs with explanations, saying, for instance, that "Murder Incorporated" was about the idea that there is "an acceptable body count for a certain portion of our citizenry." His good intentions were plain.

Most of the set drew on Mr. Springsteen's new album, "The Ghost of Tom Joad" (Columbia); he started with the title song, in which he searches for the character from "The Grapes of Wrath" who said he'll be there waiting "wherever somebody's strugglin' to be free." He also revived cheerless older songs like "Adam Raised a Cain" and "Reason to Believe," tapping the pessimistic streak that has been growing in his repertory for 20 years. Where he once saw open highways, he now sees roads to nowhere.

At the Beacon, the music was sombre and austere, with flashes of bitter vehemence. Mr. Springsteen sang with a gruff, burdened voice, moving deliberately through his cramped melodies as if he was giving mournful testimony.

Questions

1. What word best describes Mr. Springsteen's music?

    (1) Joyous

    (2) Content

    (3) Edgy

    (4) Hopeless

2. What do most of Mr. Springsteen's songs say about life?

    (1) It's confusing.

    (2) It's fun.

    (3) It's unfair.

    (4) It's mysterious.

3. Springsteen's performance of these songs was

    (1) believable.

    (2) disappointing.

    (3) lively.

    (4) humorous.

4. What do you think Pareles means when he says "Where he [Springsteen] once saw open highways, he now sees roads to nowhere"?

    (1) Springsteen used to be shy, but now he's outgoing.

    (2) Springsteen used to be humorous, but now he's serious.

    (3) Springsteen used to be optimistic, but now he's pessimistic.

    (4) Springsteen used to take risks, but now he's cautious.

Answers

1. According to the passage, the word that would best describe Springsteen's music is (4), hopeless. This is clear from the content of Springsteen's songs ("about people crushed by hard times and bad decisions"), and from the reviewer's diction: he calls Springsteen's old songs "cheerless," talks of Springsteen's "pessimistic streak," calls the performance "sombre," and speaks of Springsteen's "burdened voice," and his "mournful testimony."

2. Most of Springsteen's songs say that life is (3), unfair. We can infer this from the first paragraph, where the reviewer says Springsteen sang "about people crushed by hard times and bad decisions" and from Springsteen's explanation for "Murder Incorporated," which he said "was about the idea that there is an 'acceptable body count for a certain portion of our citizenry.'"

3. Springsteen's performance of these songs was, according to the reviewer, (1), believable. It certainly wasn't (3), lively or (4), humorous. And because the reviewer tells us that Springsteen's "good intentions were plain," that the music was "sombre and austere," and that Springsteen sang "with a gruff, burdened voice...as if he was giving mournful testimony" (evidence), we can say that Springsteen's performance was believable.

4. When Pareles says "Where he once saw open highways, he now sees roads to nowhere," he means (3), Springsteen used to be optimistic, but now he's pessimistic. Pareles mentions that in this performance Springsteen capped "the pessimistic streak that has been growing in his repertory for 20 years."

MOVIE AND TV REVIEWS

Because movie making has become such a booming industry, reviewers of films have a lot to look for. They will consider the quality of the acting, casting, costume design, music, and editing. They will look at the movie's themes, its believability, its pace, and the originality of the script. Essentially a movie reviewer must take into consideration all of the elements of drama as well as the technical elements that make up cinematography, such as lighting, camera angles, focus, and so on.

TV reviewers have much the same task, but their job is not quite as complicated. Because television programs are usually much shorter and have much smaller budgets than movies, there are lower expectations. A key element to television reviews is the consideration of the audience. Who will be viewing this show which airs at such and such a time? How does it compare to other shows aired at the same time? What makes this show different from others?

Below are excerpts from two movie reviews, both by the same critic, Janet Maslin. One is quite positive, the other negative. As you read, be sure to distinguish between the facts Maslin provides and her opinions. Also, watch how Maslin supports her opinions and how she chooses words that express her feelings about the film.

TOY STORY

Raised high above his humble station, Mr. Potato Head is now movie royalty, a star of the sweetest and savviest film of the year. The computer-animated "Toy Story," a parent-tickling delight, is a work of incredible cleverness in the best two-tiered Disney tradition. Children will enjoy a new take on the irresistible idea of toys coming to life. Adults will marvel at a witty script and utterly brilliant anthropomorphism. And maybe no one will even mind what is bound to be a mind-boggling marketing blitz. After all, the toy tie-ins are two old friends.

It's a lovely joke that the film's toy characters are charmingly plain (Etch-a-Sketch, plastic soldiers, a dog made out of a Slinky) while its behind-the-scenes technology, under the inspired direction of John Lasseter, could not be more cutting edge. It's another joke that this film begins with human characters who have the flat, inexpressive look of toys. A boy named Andy is seen playing boisterously with Woody, his favorite cowboy, whose features remain innocently blank. Only after Andy gets bored and goes elsewhere does Woody spring magically to life.

Questions

1. How can you tell that Maslin likes this movie? Cite specific words and phrases.

2. Why does Maslin like this movie? What reasons does she provide?

3. What facts does Maslin provide about the film?

4. Would you go see this movie? Why or why not?

Answers

1. We can tell that Maslin likes this movie from her first sentence, in which she calls "Toy Story" the "sweetest and savviest film of the year." She also calls it a "parent-tickling delight" and a "work of incredible cleverness."

2. She likes the movie for several reasons. First, the movie will appeal to parents as well as to children. (Parents will particularly enjoy the "witty script.") Second, the story is a "new take" on a familiar storyline, so it is original. Third, the toys who star in the movie are "old friends" and "charmingly plain" while the technology used to create the film "could not be more cutting edge." And fourth, the toys are more human than the people.

3. Maslin provides a few facts about the film. One of its toy stars is Mr. Potato Head, it is computer-animated, and it is directed by John Lasseter. The boy who owns the toys is named Andy, and Woody is Andy's favorite cowboy. When Andy leaves the room, Woody comes to life.

4. Answers will vary.

NICK OF TIME

In "Nick of Time," Johnny Depp plays an ordinary man with an extraordinary problem: total strangers have kidnapped his daughter, given him a gun and insisted that he kill an assigned target, who happens to be the Governor of California. And he must accomplish this in little more than an hour, which means that the film is loaded with shots of clocks ticking off minutes. Despite such constant evidence of time passing, you will be tempted to look at your watch. Often.

"Nick of Time" calls for Hitchcockian cleverness to make the real-time device work (and Hitchcock had his own troubles with it in "Rope"). Instead, John Badham supplies pumped-up direction that often seems needlessly feverish, and plenty of little touches that don't quite pay off. "I'm not gonna give you no key," a hotel chambermaid tells Mr. Depp, who is trying to break into the Governor's room. But then she drops the key by accident. The film isn't often more ingenious than that.

Questions

1. How can you tell that Maslin dislikes this movie? Cite specific words and phrases.

2. Why does Maslin dislike this movie? What reasons does she provide?

3. What facts does Maslin provide about the film?

4. Would you go see this movie? Why or why not?

Answers

1. The first evidence that Maslin dislikes this movie are the two sentences that end the first paragraph: "Despite such constant evidence of time passing, you will be tempted to look at your watch. Often." Maslin, it seems, couldn't wait for this movie to be over. In addition, she sees the direction "pumped-up" and "needlessly feverish," with "little touches that don't quite pay off"

2. Maslin dislikes this movie mostly because of the direction, which, she says, is not very clever. As in the scene with the chambermaid, the movie feels contrived. This type of movie, Maslin says, requires a type of "Hitchcockian cleverness" which even Hitchcock himself had trouble with, and she clearly doesn't think the director, John Badham, comes close to achieving that cleverness.

3. Maslin tells us the basic plot of the film: Johnny Depp stars as a man whose daughter has been kidnapped, and in order to get her back, he must kill the Governor of California—all within an hour. She also tells us the name of the director and about a particular scene later in the film. We also know that there are many shots of clocks counting down the time Depp's character has left.

4. Answers will vary.

Critics, then, provide a valuable service by "screening" works of art for us. Good commentary is also helpful in that it often makes us more aware of what qualities make a successful work of art. We must be careful, though, to remember that this is only one person's opinion. Two different commentators may have completely different opinions about a movie or a book, and our opinion may be different from both of theirs.

Practice: Applying Comprehension Strategies

DIRECTIONS: Read the following passages. Apply the four reading strategies you've learned as well as your vocabulary skills. Write in the margins and mark up the text as you go. Then answer the questions following each passage.

Passage A—Book Review

"Fatal Attraction" by Tina Rosenberg. Review of: *The Stalking of Kristin: A Father Investigates the Murder of his Daughter,* by George Lardner Jr.

On May 30, 1992, Kristin Lardner, a 21-year-old college student, was stalked and shot to death in Boston by her abusive former boyfriend, Michael Cartier, who then killed himself. Kristin's father, George Lardner Jr., an investigative reporter at The Washington Post, asked his editors to let him write about her case. Kristin's murder, and Mr. Lardner's reporting in The Post, spurred Massachusetts to toughen the judicial system's handling of batterers. "The Stalking of Kristin" entwines Mr. Lardner's Pulitzer Prize-winning reportage on the case with the story of his daughter's life. The result, unfortunately, is a wobbly and at times tedious book, but the deadly negligence it exposes demands attention.

[...]

Mr. Lardner is not just an investigative reporter; he is a man possessed. He talked to everyone with a detail to contribute, including Cartier's parents. He confronted one court official after another with specifics about their negligence, his unique dual role allowing him to act out the fantasy, shared by most crime victims, of shaking the system's bureaucrats by the collar until they understand the anguish they have caused. But the book suffers. An author's passion can make a book about a wrenching personal experience memorable, but without a careful editor to keep it in perspective, it can be ruinous.

[...]

But the book's most important lesson is that Kristin did everything right. Mr. Lardner refutes the widespread belief that the courts offer effective protection to battered women, and that only women who fail to report domestic violence or drop charges continue to fall victim. Unlike many battered women, Kristin was educated, sophisticated, and free of the need to worry about children, with the time and resources to make the law work for her. Most important, she was a member of the class of people who believe the law when it promises to protect them. "The Stalking of Kristin" reveals the tragic error of that trust.

1. Why is Lardner's book not as good as his reporting about his daughter's death? What makes his book "suffer"?

2. What about Lardner's book is positive?

3. How did Lardner's reporting benefit citizens?

4. Why is Kristin a special case of a battered woman?

5. In your own words, what is the book's most important lesson?

Passage B—Music Review

"Underflowers' Music Shuns Screaming Angst," by Ed Masley (Copyright, Pittsburgh Post-Gazette. Reprinted with Permission.)

Night closes in on a cool October sky as members of Underflowers gather around a bite-size table in the South Side Tuscany coffee shop, quietly discussing the decidedly de-caf magic of their haunting new self-titled CD.

At one point, guitarist and some-time vocalist J Orazi refers to it in passing as *calming music*.

"Not always rock 'n' roll," he says.

To tell you the truth, it's hard to imagine Underflowers as even *occasionally* rock 'n' roll.

They have no drummer.

The electric guitar is used more for shading than banging about.

And as violinist Tanya Kavalkovich is keen to point out, the flier for one recent D.C. appearance did, in fact, urge fans to *come relax*.

[...]

Underflowers never screams, regardless of what kind of day Rollins is having. On tracks like "Aurora Ring," "Inn" and "For Luna," Chesla's vocals was over the mix with such delicate, spellbinding grace and beauty, you'd swear there were angels at play in the speakers.

1. Is the word *haunting* in the first paragraph negative or positive?

2. Why are Underflowers hard to imagine as rock 'n' roll?

3. How does Masley feel about the band's vocalist?

4. Underflowers music can be best described as:

(1) furious.

(2) mesmerizing.

(3) heavy.

(4) simple.

(5) frightening.

Answers

Passage A—Book Review

1. According to Rosenberg, Lardner's book is not as good as his reporting because "he is a man possessed." The book suffers because it is too passionate about his subject, and he has not edited it enough to "keep it in perspective."

2. The book is successful in that it "reveals the tragic error" of citizens like Kristin who trust the law when it says that it will protect them. The "deadly negligence" on the part of our judicial system that is exposed by the book "demands attention." Despite its faults, the book addresses a very important issue.

3. Lardner's reporting benefited citizens of Massachusetts by persuading Massachusetts lawmakers to toughen their laws regarding battered women.

4. Kristin is a special case because unlike many battered women, she "did everything right." She was also more "educated, sophisticated," liberated, and privileged than most battered women. And yet nothing could protect her.

5. Answers will vary.

Passage B—Music Review

1. The word "haunting" is positive. This is evident from the other positive terms used in the passage, such as "calming," "relax," "delicate," and especially "spellbinding grace and beauty."

2. Underflowers are hard to imagine as rock'n'roll because there is a violinist instead of a drummer, the electric guitar is subtle instead of screaming, and their music aims to relax their listeners.

3. Masley finds the vocalist's voice mesmerizing. He calls it "delicate," and says she sings with "spellbinding grace and beauty." He compares her voice to "angels at play in the speakers." This is quite a high compliment.

4. Underflowers music can best be described as (2), mesmerizing.

REVIEW

Commentary is literature written to illuminate or explain other works of art and literature. Commentary lets us know how critics feel about a work and why they feel the way they do.

When reading commentary, it is particularly important to distinguish between fact and opinion so that while we might appreciate a critic's judgement, we are still able to draw conclusions of our own.

There are many types of commentary. Four types of reviews we discussed were book, theater, music, and movie/TV reviews. In each category, critics examine how well the artists perform within their respective genres. One of the key elements critics look for is originality.

Interpreting Literature and the Arts

Post-Test

INTERPRETING LITERATURE AND THE ARTS

POST-TEST

> **DIRECTIONS:** Read each of the passages below and then answer the questions pertaining to them. Choose the <u>best answer choice</u> for each question.

Questions 1–9 are based on the following passage

PASSAGE A
THOSE WINTER SUNDAYS

(1) Sundays too my father got up early
(2) and put his clothes on in the blueblack cold,
(3) then with cracked hands that ached
(4) from labor in the weekday weather made
(5) banked fires blaze. No one ever thanked him.
(6) I'd wake and hear the cold splintering, breaking,
(7) When the rooms were warm, he'd call,
(8) and slowly I would rise and dress,
(9) fearing the chronic angers of that house,
(10) Speaking indifferently to him,
(11) who had driven out the cold
(12) and polished my good shoes as well.
(13) What did I know, what did I know
(14) of love's austere and lonely offices?

—By Robert Hayden

1. "Sundays too" tells us that

    (1) the speaker's father never got up early.
    (2) the speaker's father got up early Mondays through Saturdays.
    (3) the speaker's father didn't work.
    (4) the speaker's father had to work on Sundays.
    (5) the father didn't work on Sundays.

2. The speaker's father got up early on Sundays

    (1) to go to work.
    (2) to be thanked.
    (3) because he was angry.
    (4) to warm the house.
    (5) to go into town.

3. No one ever thanked the speaker's father because

    (1) they were afraid of him.
    (2) they were too cold.
    (3) they were lonely.
    (4) he was already off to work.
    (5) they didn't care.

4. The speaker's father could be

    (1) a doctor.
    (2) a lawyer.

(3) a manual laborer.

(4) a dentist.

(5) unemployed.

5. The theme of this poem could be stated as

   (1) love is blind.

   (2) we don't always know love when we see it.

   (3) it is better to have loved and lost than never to have loved at all.

   (4) we all need to be loved.

   (5) you don't appreciate what you have until its gone.

6. We can infer that the speaker now

   (1) hates his father.

   (2) is afraid of his father.

   (3) lives with his father.

   (4) understands his father.

   (5) sees his father regularly

7. We can infer that the speaker

   (1) thanked his father.

   (2) wishes he had thanked his father.

   (3) wishes he hadn't thanked his father.

   (4) thinks his father should thank him.

   (5) doesn't want to thank his father.

8. The speaker's father

   (1) loved his children.

   (2) beat his children.

   (3) hated his children.

   (4) never spoke to his children.

   (5) abandoned his children.

9. Line 6 in the poem uses

   (1) irony.

   (2) alliteration.

   (3) metaphor.

   (4) simile.

   (5) paradox.

Questions 10–15 are based on the following passage.

PASSAGE B

In my opinion the most profitable and most natural exercise of our mind is *conversation*. To me it is a more agreeable oc-cupation than any other in life; and for that
(5) reason, if I were at this moment obliged to choose, I would sooner consent, I think, to lose my sight than my hearing or speech. The Athenians, and still more the Romans, held this practice in great honour
(10) in their Academies. To this day the Italians preserve some traces of it, and greatly to their benefit, as may be seen if we com-pare ourselves with them in intelligence.

The study of books is a feeble and
(15) languid action which does not warm us, whilst conversation instructs and exercises us at the same time. If I converse with a man of strong mind and a stiff jouster, he will press on my flanks, prick me to right
(20) and left; his ideas will give an impetus to mine. Rivalry, vain-glory, strife, stimulate me and lift me above myself. And agree-ment is an altogether tiresome element in conversation.

(25) As our mind is strengthened by com-munication with vigorous and well regu-lated minds, it is not to be imagined how much it looses and deteriorates by con-tinual intercourse and association with vul-
(30) gar and feeble-minded people. There is no infection which spreads like that. I know

well enough by experience how much a yard it costs.

From "Of the Art of Conversing"
by Michael DeMontaigne

10. The author would rather

    (1) be blind than deaf.
    (2) be mute than blind.
    (3) be deaf than mute.
    (4) be paralyzed than blind.
    (5) be blind then paralyzed.

11. The author likes conversations that are

    (1) without conflict or confrontation.
    (2) full of vibrant debate.
    (3) one-sided.
    (4) humorous.
    (5) brief.

12. The author thinks speaking with "vulgar and feeble-minded people"

    (1) improves the mind.
    (2) infects the mind.
    (3) regulates the mind.
    (4) helps you sympathize with them.
    (5) weakens the mind.

13. The author would rather

    (1) read than talk.
    (2) write than read.
    (3) argue than read.
    (4) write than argue.
    (5) read than argue.

14. Conversation is "most profitable" because

    (1) it improves the mind.
    (2) it improves vocabulary.
    (3) it is honorable.
    (4) it is good exercise.
    (5) it improves social status.

15. A good conversation will

    (1) teach.
    (2) stimulate.
    (3) strengthen.
    (4) All of the above.
    (5) None of the above.

Questions 16–21 are based on the following excerpt.

PASSAGE C

TOM: —Oh.—Laura…

AMANDA: (*touching his sleeve*) You know how Laura is. So quiet but—still water runs deep! She notices things and I think she—broods about them. (*Tom looks up.*) A few days ago I came in and she was crying.

TOM: What about?

AMANDA: You.

TOM: Me?

AMANDA: She has an idea that you're not happy here.

TOM: What gave her that idea?

AMANDA: What gives her any idea? However, you do act strangely. I—I'm not criticizing, understand *that*! I know your ambitions do not lie in the warehouse, that like everybody in the whole wide world—you've had to—make sacrifices, but—Tom—Tom—life's not easy, it calls for—Spartan endurance! There's so many things in my heart that I cannot describe to you! I've never told you but I—*loved* your father....

TOM: (*gently*) I know that, Mother.

AMANDA: And you—when I see you taking after his ways! Staying out late—and—well, you *had* been drinking the night you were in that—terrifying condition! Laura says that you hate the apartment and that you go out nights to get away from it! Is that true, Tom?

TOM: No. You say there's so much in your heart that you can't describe to me. That's true of me, too. There's so much in my heart that I can't describe to *you*! So let's respect each other's—

AMANDA: But, why—*why*, Tom—are you always so *restless*? Where do you go to, nights?

TOM: I—go to the movies.

AMANDA: Why do you go to the movies so much, Tom?

TOM: I go to the movies because—I like adventure. Adventure is something I don't have much of at work, so I go to the movies.

AMANDA: But, Tom, you go to the movies *entirely too much*!

TOM: I like a lot of adventure.

Amanda looks baffled, then hurt. As the familiar inquisition resumes he becomes hard and impatient again. Amanda slips back into her querulous attitude toward him.

(*Image on screen: Sailing vessel with Jolly Roger.*)

AMANDA: Most young men find adventure in their careers.

TOM: Then most young men are not employed in a warehouse.

AMANDA: The world is full of young men employed in warehouses and offices and factories.

TOM: Do all of them find adventure in their careers?

AMANDA: They do or they do without it! Not everybody has a craze for adventure.

TOM: Man is by instinct a lover, a hunter, a fighter, and none of those instincts are given much play at the warehouse!

AMANDA: Man is by instinct! Don't quote instinct to me! Instinct is something that people have got away from! It belongs to animals! Christian adults don't want it!

TOM: What do Christian adults want, then, Mother?

AMANDA: Superior things! Things of the mind and the spirit! Only animals have to satisfy instincts! Surely your aims are somewhat higher than theirs! Than monkeys—pigs—

TOM: I reckon they're not.

From "The Glass Menagerie," by Tennessee Williams

16. Why is Laura upset?

   (1) Tom has left home.
   (2) Tom goes to the movies without her.
   (3) She thinks Tom is unhappy.
   (4) Tom cannot describe his feelings.
   (5) Tom is happy with her.

17. Tom goes to the movies

   (1) because he wants to be an actor.
   (2) because he likes to be alone.
   (3) because he is Christian.
   (4) because he is restless.
   (5) because he wants to be away from Amanda.

18. Tom's job

    (1) is very exciting.

    (2) pays well.

    (3) is a night job.

    (4) keeps him happy.

    (5) lacks excitement.

19. Amanda is upset because

    (1) Tom wants to quit his job.

    (2) Tom is always out.

    (3) Tom made Laura cry.

    (4) Tom is a pig.

    (5) Tom is unfaithful.

20. We can tell from this passage that Tom and Amanda

    (1) are very open with each other.

    (2) fight often.

    (3) are adventurous.

    (4) are happy.

    (5) agree on everything.

21. Amanda thinks Tom

    (1) should go out more.

    (2) should get a better job.

    (3) should apologize to his sister.

    (4) should be satisfied with what he has.

    (5) should take her to the movies.

Questions 22–29 are based on the following passage.

PASSAGE D

"Quick! Jump for the woods!"

We done it, and then peeped down the woods through the leaves. Pretty soon a splendid young man came galloping down
(5) the road, setting his horse easy and looking like a soldier. He had his gun across his pommel. I had seen him before. It was young Harney Shepherdson. I heard Buck's gun go off at my ear, and Harney's
(10) hat tumbled off from his head. He grabbed his gun and rode straight to the place where we was hid. But we didn't wait. We started through the woods on a run. The woods warn't thick, so I looked over my
(15) shoulder to dodge the bullet, and twice I seen Harney cover Buck with his gun; and then he rode away the way he come—to get his hat, I reckon, but I couldn't see. We never stopped running till we got home.
(20) The old gentleman's eyes blazed a minute—'twas pleasure, mainly, I judged—then his face sort of smoothed down, and he says, kind of gentle:

"I don't like that shooting from be-
(25) hind a bush. Why didn't you step into the road, my boy?"

"The Shepherdsons don't, father. They always take advantage."

Miss Charlotte she held her head up
(30) like a queen while Buck was telling his tale, and her nostrils spread and her eyes snapped. The two young men looked dark, but never said nothing. Miss Sophia she turned pale, but the color come back when
(35) she found the man warn't hurt.

Soon as I could get Buck down by the corn-cribs under the trees by ourselves, I says:

"Did you want to kill him, Buck?"
(40) "Well, I bet I did."

"What did he do to you?"

"Him? He never done nothing to me."

"Well, then, what did you want to kill him for?"

(45) "Why, nothing—only it's on account of the feud."

"What's a feud?"

"Why, where was you raised? Don't you know what a feud is?"

(50) "Never heard of it before—tell me about it."

"Well," says Buck, "a feud is this way: A man has a quarrel with another man, and kills him; then that other man's (55) brother kills *him*; then the other brothers, on both sides, goes for one another; then the *cousins* chip in—and by and by everybody's killed off, and there ain't no more feud. But it's kind of slow, and takes (60) a long time."

"Has this one been going on long, Buck?"

"Well, I should *reckon!* It started thirty year ago, or som'ers along there. (65) There was trouble 'bout something, and then a lawsuit to settle it; and the suit went agin one of the men, and so he up and shot the man that won the suit—which he would naturally do, of course. Anybody (70) would."

"What was the trouble about, Buck?—land?"

"I reckon maybe—I don't know."

"Well, who done the shooting? Was (75) it a Grangeford or a Shepherdson?"

"Laws, how do *I* know? It was so long ago."

"Don't anybody know?"

"Oh, yes, pa knows, I reckon, and (80) some of the other old people; but they don't know now what the row was about in the first place."

From *The Adventures of Huckleberry Finn,* by Mark Twain

22. Buck's full name is

(1) Buck Finn.

(2) Buck Shepherdson.

(3) Buck Grangeford.

(4) Buck Twain.

(5) Mark Buckwald.

23. Buck shoots at Harney because

(1) Harney shot at him.

(2) Harney insulted Buck.

(3) Harney is a Shepherdson.

(4) Harney had a gun.

(5) He thought Harney was someone else.

24. Miss Sophia turned pale because

(1) she was afraid Harney got hurt.

(2) she hates Harney.

(3) she was afraid Buck got hurt.

(4) she was angry that Buck shot from behind a bush.

(5) she was afraid she would get shot.

25. Buck wanted to kill Harney

(1) for his money.

(2) because of a feud between their families.

(3) because Harney was suing him.

(4) because they were cousins.

(5) because Harney stole his wife.

26. The narrator's language shows that

(1) he is well educated.

(2) he is a foreigner.

(3) he is an orphan.

(4) he has had little schooling.

(5) he is a politician.

27. The families started fighting over

    (1) a woman.

    (2) land.

    (3) money.

    (4) a trespassing incident.

    (5) No one knows.

28. This passage shows

    (1) the stupidity of the feud.

    (2) the importance of family loyalty.

    (3) you should never let your guard down.

    (4) buck needs more shooting practice.

    (5) some things are worth fighting for.

29. In the description of Miss Charlotte, "held her head up like a queen" is which type of figurative language?

    (1) Metaphor

    (2) Simile

    (3) Personification

    (4) Allegory

    (5) Irony

Questions 30–35 are based on the following poem.

PASSAGE E
RICHARD CORY

(1) Whenever Richard Cory went down town,
(2) We people on the pavement looked at him:
(3) He was a gentleman from sole to crown,
(4) Clean favored, and imperially slim.
(5) And he was always quietly arrayed,
(6) And he was always human when he talked;
(7) But still he fluttered pulses when he said,
(8) "Good-morning," and he glittered when he walked.
(9) And he was rich—yes, richer than a king—
(10) And admirably schooled in every grace:
(11) In fine, we thought that he was everything
(12) To make us wish that we were in his place.
(13) So on we worked, and waited for the light,
(14) And went without meat, and cursed the bread;
(15) And Richard Cory, one calm summer night,
(16) Went home and put a bullet through his head.

By Edwin Arlington Robinson "Richard Cory."

30. Richard Cory was

    (1) hated by everyone.

    (2) loved by everyone.

    (3) envied by everyone.

    (4) looked down upon by everyone.

    (5) ignored by everyone.

31. The "we" in this poem refers to

    (1) members of the upper class.

    (2) members of the royalty.

(3) slaves.

(4) members of the working class.

(5) members of society as a whole.

32. In line 13, "waited for the light" suggests

    (1) they were waiting to be enlightened.

    (2) they were waiting for electricity.

    (3) they were waiting for bread.

    (4) they were waiting for Richard Cory.

    (5) they were waiting for riches.

33. Richard Cory shot himself because

    (1) he had lost his money.

    (2) he was cursed by the people.

    (3) he had money but not happiness.

    (4) he had meat but no bread.

    (5) it was an accident.

34. This poem is

    (1) a sonnet.

    (2) a ballad.

    (3) free verse.

    (4) a limerick.

    (5) a haiku.

35. The theme of this poem is

    (1) money is the root of all evil.

    (2) the more you have, the more you want.

    (3) money can't buy you happiness.

    (4) money makes the world go around.

    (5) a lot of money will make you unhappy.

Questions 36–40 are based on the following passage.

PASSAGE F

Of all the stories in *A Curtain of Green*, "Lily Daw and the Three Ladies" represents most clearly the choice open to white southern women in [Eudora] Welty's
(5) early stories. (Perhaps it is for this reason that Welty chose it to open her first collection.) That choice is suggested by the story's title: a woman may either be a married and respectable lady, like Mrs. Carter
(10) and her friend Mrs. Watts, or she may be eccentric like the retarded girl Lily Daw, continually threatened by scandal, madness, and confinement.

Lily grew up as a ward of the ladies
(15) of Victory, Mississippi, but now, in her teens, she has suddenly begun acting independently, talking back to her elders, sneaking away to circuses, and showing an interest in the opposite sex. The town's
(20) matriarchs first decide that confinement in an asylum can be the only remedy, but then, in a comic reversal caused partly by Lily's boyfriend's faithfulness and partly by the rebellion of one of their own mem-
(25) bers, Aimee Slocum, they decide that Lily's marrying her boyfriend must be accepted after all. By juxtaposing marriage and madness in such an exaggerated way, Welty's story comically subverts the ladies'
(30) authority. Their ideal of marriage comes to seem like a kind of confining madness, whereas Lily's unconscious defiance of the community's standards becomes a liberating sanity, an escape and transformation.
(35) Yet even this twist is not the last the story has in store for us. Our euphoria over Lily's apparent victory over the ladies fades slightly when the story is reread: even as she is united with her lover we can
(40) see the ladies extending their control over her again. The story's acerbity and exuber-

ance hardly suggests anxiety on Welty's part, yet its pointed linking of rebelliousness and madness provides an appropriate
(45) entry into the other, darker stories in *A Curtain of Green* about the potentially disastrous consequences of nonconformity.

> From "Lily Daw and the Three Ladies"
> in *The Heart of the Story,*
> *Eudora Welty's Short Fiction*

36. "Lily Daw and the Three Ladies" is the first story in *A Curtain of Green* because

    (1) it is darker than the other stories.
    (2) it clearly establishes the theme of the other stories.
    (3) it is the best story.
    (4) it is the longest story.
    (5) it was written before the others.

37. Lily is nearly confined in an asylum because

    (1) she begins to rebel.
    (2) she joins the circus.
    (3) she is insane.
    (4) she gets pregnant.
    (5) she's using drugs.

38. The irony of this story is that

    (1) Lily is really mad after all.
    (2) Aimee Slocum rebels too.
    (3) it is the first in the collection.
    (4) Lily is more sane than the ladies.
    (5) the ladies were right about her.

39. The message of the story appears to be

    (1) it is best to conform.
    (2) rebelling doesn't pay.
    (3) Lily should have had a mother to raise her properly.
    (4) marriage is madness.
    (5) women should not be forced to conform.

40. The author of this passage thinks Welty

    (1) is a gifted writer.
    (2) is a terrible writer.
    (3) writes too many stories about the same thing.
    (4) is a funny writer.
    (5) is limited in her ability.

Questions 41–46 refer to the following passage.

PASSAGE G

I had halted on the road. As soon as I saw the elephant I knew with perfect certainty that I ought not to shoot him. It is a serious matter to shoot a working el-
(5) ephant—it is comparable to destroying a huge and costly piece of machinery—and obviously one ought not to do it if it can possibly be avoided. And at that distance, peacefully eating, the elephant looked no
(10) more dangerous than a cow. I thought then and I think now that his attack of "must" was already passing off; in which case he would merely wander harmlessly about until the mahout came back and caught
(15) him. Moreover, I did not in the least want to shoot him. I decided that I would watch him for a little while to make sure that he did not turn savage again, and then go home.
(20) But at that moment I glanced round at the crowd that had followed me. It was

an immense crowd, two thousand at the least and growing every minute. It blocked the road for a long distance on either side.
(25) I looked at the sea of yellow faces above the garish clothes—faces all happy and excited over this bit of fun, all certain that the elephant was going to be shot. They were watching me as they would watch a
(30) conjurer about to perform a trick. They did not like me, but with the magical rifle in my hands I was momentarily worth watching. And suddenly I realized that I should have to shoot the elephant after all. The
(31) people expected it of me and I had got to do it; I could feel their two thousand wills pressing me forward, irresistibly. And it was at this moment, as I stood there with the rifle in my hands, that I first grasped
(35) the hollowness, the futility of the white man's dominion in the East. Here was I, the white man with his gun, standing in front of the unarmed native crowd—seemingly the leading actor of the piece; but in
(40) reality I was only an absurd puppet pushed to and fro by the will of those yellow faces behind. I perceived in this moment that when the white man turns tyrant it is his own freedom that he destroys.

From "Shooting an Elephant," by George Orwell.

41. The author shouldn't shoot the elephant because

    (1) he is not the owner of the elephant.
    (2) it is illegal to shoot elephants.
    (3) the elephant is a valuable worker.
    (4) the elephant was no longer able to work.
    (5) elephants were an endangered species.

42. The author was chasing the elephant because

    (1) it had been acting savagely.
    (2) it belonged to him and had run away.
    (3) he was hunting.
    (4) it was blocking the road.
    (5) it had escaped from the zoo.

43. A "conjurer" is

    (1) a native.
    (2) a circus ringmaster.
    (3) a murderer.
    (4) a magician.
    (5) a psychic.

44. The author shot the elephant because

    (1) it was savage and dangerous.
    (2) he wanted to.
    (3) the people wanted him to.
    (4) he was ordered to.
    (5) he would be killed if he didn't.

45. This event took place in

    (1) America.
    (2) Asia.
    (3) Africa.
    (4) Antarctica.
    (5) Europe.

INTERPRETING LITERATURE AND THE ARTS

ANSWER KEY

1. (2)	13. (3)	25. (2)	37. (1)
2. (4)	14. (1)	26. (4)	38. (4)
3. (1)	15. (4)	27. (5)	39. (5)
4. (3)	16. (3)	28. (1)	40. (1)
5. (2)	17. (4)	29. (2)	41. (3)
6. (4)	18. (5)	30. (3)	42. (1)
7. (2)	19. (2)	31. (4)	43. (4)
8. (1)	20. (2)	32. (5)	44. (3)
9. (3)	21. (4)	33. (3)	45. (2)
10. (1)	22. (3)	34. (2)	
11. (2)	23. (3)	35. (3)	
12. (5)	24. (1)	36. (2)	

Pre-GED Interpreting Literature and the Arts

POST-TEST SELF-EVALUATION

Question Number	Subject Matter Tested	Section to Study (section, heading)
1.	Plot in poetry	IV, Reading Poetry, Elements of Sense
2.	Plot in poetry	IV, Reading Poetry, Elements of Sense
3.	Plot in poetry	IV, Reading Poetry, Elements of Sense
4.	Inference, plot in poetry	II, From Active Reading to Comprehension, and IV, Reading Poetry, Elements of Sense
5.	Theme	III, Elements of Fiction, and IV, Reading Poetry, Elements of Sense
6.	Theme	III, Elements of Fiction, and IV, Reading Poetry, Elements of Sense
7.	Theme	III, Elements of Fiction, and IV, Reading Poetry, Elements of Sense
8.	Theme	III, Elements of Fiction, and IV, Reading Poetry, Elements of Sense
9.	Figurative Language	III, Elements of Fiction, and IV, Reading Poetry, Elements of Sense
10.	Essay, general comprehension	II, Active Reading, and III, The Essay
11.	Essay, general comprehension	II, Active Reading, and III, The Essay
12.	Essay, general comprehension	II, Active Reading, and III, The Essay
13.	Essay, general comprehension	II, Active Reading, and III, The Essay
14.	Essay, main idea	III, The Essay
15.	Essay, main idea	III, The Essay
16.	Plot in drama	III, Elements of Fiction, and V, Elements of Drama
17.	Understanding character through dialogue	V, Elements of Drama
18.	Understanding character through dialogue	V, Elements of Drama
19.	Understanding character through dialogue	V, Elements of Drama
20.	Understanding character through dialogue	V, Elements of Drama
21.	Understanding character through dialogue	V, Elements of Drama
22.	Identifying details	II, Active Reading
23.	Understanding motive in fiction	III, Elements of Fiction
24.	Inference	II, From Active Reading to Comprehension

II = Reading Literature   III = Reading Prose   IV = Reading Poetry   V = Reading Drama   VI = Reading Commentary

Question Number	Subject Matter Tested	Section to Study (section, heading)
25.	Understanding motive in fiction	III, Elements of Fiction
26.	Identifying details	II, Active Reading
27.	Plot in Fiction	III, Elements of Fiction
28.	Theme	III, Elements of Fiction, and IV, Reading Poetry, Elements of Sense
29.	Figurative Language	III, Elements of Fiction, and IV, Reading Poetry, Elements of Sense
30.	Inference, Plot in Poetry	II, From Active Reading to Comprehension, and IV, Reading Poetry, Elements of Sense
31.	Inference, Plot in Poetry	II, From Active Reading to Comprehension, and IV, Reading Poetry, Elements of Sense
32.	Figurative Language	III, Elements of Fiction, and IV, Reading Poetry, Elements of Sense
33.	Inference, Plot in Poetry	II, From Active Reading to Comprehension, and IV, Reading Poetry, Elements of Sense
34.	Poetic Structure	IV, Reading Poetry, Poetic Forms
35.	Theme	III, Elements of Fiction, and IV, Reading Poetry, Elements of Sense
36.	Inference	II, From Active Reading to Comprehension
37.	Plot in Fiction	III, Elements of Fiction
38.	Figurative Language, Irony	III, Elements of Fiction, and V, Elements of Drama
39.	Theme	III, Elements of Fiction, and IV, Reading Poetry, Elements of Sense
40.	Inference	II, From Active Reading to Comprehension
41.	Inference	II, From Active Reading to Comprehension
42.	Plot in Fiction	III, Elements of Fiction
43.	Vocabulary in Context	II, Active Reading
44.	Plot in Fiction	III, Elements of Fiction
45.	Identifying Details	II, Active Reading

II = Reading Literature   III = Reading Prose   IV = Reading Poetry   V = Reading Drama   VI = Reading Commentary

POST-TEST ANSWERS AND EXPLANATIONS

Passage A

1. **(2)** "Sundays too" tells us that (2), the speaker's father got up early Mondays through Saturdays. We can't be sure that (4), the speaker's father had to work on Sundays, the poem only tells us that he got up early. We know answer (1), the speaker's father never got up early, cannot be correct because line 1 tells us that he did get up early on Sundays. Finally, answers (3) and (5) can't be correct, because the poem doesn't provide any evidence that the speaker's father didn't work. In fact, it suggests the opposite: that the speaker's father was a hard-working man.

2. **(4)** The speaker's father got up early on Sundays to (4), warm the house. We can tell this is the answer because he wouldn't wake the others until the house was warm (line 7). Also, lines 4–5 tell us that that was the first thing he did after he put on his clothes. We don't know whether or not the father had to work (1), or whether he wished to be thanked (2), or if he was angry (3) (we only know that the speaker was afraid of the anger in the house). We don't know if the father went into town on Sundays (5) but we know this wasn't his main purpose.

3. **(1)** No one ever thanked the speaker's father because (1), they were afraid of him. This is revealed in lines 8–10: "and slowly I would rise and dress, / fearing the chronic angers of that house, / Speaking indifferently to him." We know that (2), they were too cold, cannot be correct, because the father got up in order to warm the house. We can't choose answers (3), (4), or (5) because there's no evidence in the poem that they were lonely, that the father was already off to work, or that they didn't care.

4. **(3)** The speaker's father could be (3), a manual laborer. We can make this assumption because lines 3–4 tell us that the father's hands were "cracked" and that they "ached / from labor in the weekday weather." If he were a doctor (1), a lawyer (2), or a dentist (4) he wouldn't labor outside during the week. If he were unemployed (5), he wouldn't be laboring at all.

5. **(2)** The theme of this poem could be stated as (2), we don't always know love when we see it. This is revealed in the last two lines of the poem, in which the speaker laments the fact that he didn't understand "love's austere and lonely offices"—in other words, that he didn't understand that his father was showing him love by warming the house and polishing his shoes.

6. **(4)** We can infer that the speaker now (4), understands his father. Again, the last two lines—"What did I know, what did I know / of love's austere and lonely offices?"—suggest that the speaker now *does* know. This is reinforced by the fact that "what did I know" is said not once, but twice.

7. **(2)** We can infer that the speaker (2), wishes he had thanked his father. We know that "No one ever thanked him," so (1), thanked his

father and (3), wishes he hadn't thanked his father, can't be correct. Furthermore, there is no evidence in the poem that the speaker (4), thinks his father should thank him. The poem is about what the father did for the speaker, not vice versa. (5) is incorrect because we know that the speaker would very much like to thank his father.

8. **(1)** The speaker's father (1), loved his children. This is the only answer for which the poem provides evidence. Even though the poem tells us that there was a "chronic anger" in the house, we have no evidence that the father (2), beat his children or (3) hated his children (one can be angry at one's children without hating them). We also know from line 7 that answer (4), the father never spoke to his children, can't be correct, because the father would call them when the house was warm. From that we know of the father he was very dedicated and we know he didn't abandon his children, so (5) is incorrect.

9. **(3)** Line 6 in the poem uses (3), metaphor. The line compares the cold to something solid that would splinter and break from heat, like a branch or ice. There is no alliteration (2) in this line, and there is no simile (4), for there is no comparison using "like" or "as." Also, this line is not ironic (1), so (3) is the only answer that could be correct. Also, there is no example of a paradox (5).

Passage B

10. **(1)** The author would rather (1), be blind than deaf. He tells us that "I would sooner consent, I think, to lose my sight than my hearing or speech." Thus, he would rather be blind than deaf or mute.

11. **(2)** The author likes conversations that are (2), full of vibrant debate. The author writes: "Rivalry, vain-glory, strife, stimulate me and lift me above myself," while on the other hand, a conversation without that rivalry, a conversation with someone who agrees with the author, is "altogether tiresome." Thus, (1), without conflict or confrontation, cannot be correct. The author also mentions nothing about humor (4), and if he likes conversations that are lively debates, then (3), one-sided, cannot be correct either. There is no evidence that he likes conversations to be brief.

12. **(5)** The author thinks speaking with "vulgar and feeble-minded people" (5) weakens the mind. Answer (2), infects the mind, may seem like a likely answer because the writer says "there is no infection which spreads like that." However, infection is a metaphor, comparing the weakness with illness or disease. He is not saying that speaking with feeble-minded people is infectious, but rather that it is a disease which causes rapid deterioration (of the mind).

13. **(3)** The author would rather (3), argue than read. He tells us this in the first sentence of paragraph two: "The study of books is a feeble and languid action which does not warm us, whilst conversation instructs and exercises us at the same time." Furthermore, he tells us that conversation "is a more agreeable occupation than any other in life," so we may presume that he would rather converse (argue, since he doesn't like agreeable conversations) than anything else.

14. **(1)** Conversation is "most profitable" because (1), it improves the mind. The writer tells us in the second paragraph that conversation "instructs and exercises us at the same time." He also tells us in the third paragraph that "our mind is strengthened by communication." Although conversation may indeed improve vocabulary (2), there is no mention of this benefit in the passage. The writer does mention honor, but not in the context that con-

versation is honorable (3). Instead, he tells us that the Athenians and Romans "held this practice in great honour" (great esteem). Finally, answer (4), it is good exercise, is not the best answer because the profit is the improvement of the mind, not from the exercise.

15. **(4)** A good conversation will (4), all of the above. The writer tells us that conversation "instructs," so (1), teach, is correct. The writer also tells us conversation "stimulate[s]" him, so (2) is correct. And the writer tells us that the mind is "strengthened by communication with vigorous and well regulated minds," so (3) is also correct.

Passage C

16. **(3)** Laura is upset because (3), she thinks Tom is unhappy. This is revealed near the beginning of the passage when Amanda tells Tom that Laura "has an idea that you're not happy here." Although Tom does go to the movies without her (2) and although Tom cannot describe his feelings (4), neither of these are the reason Amanda expresses. Answer (1), Tom has left home, cannot be correct because Amanda specifically says Laura "has an idea that you're not happy *here*," where they are together, and we know that Tom is Amanda's son. We can safely assume that he is still living at home, so he cannot be happy without her (5).

17. **(4)** Tom goes to the movies because (4), he is restless. This is the best answer because when Amanda asks Tom why he's "always so *restless*" and where he goes at night, he replies that he goes to the movies. We also know that he's bored at work, so he goes to the movies for adventure. He doesn't go because he wants to be an actor (1)—there's no mention of this anywhere in the passage—or because he likes to be alone (2)—again, there's no mention of this, although it may indeed be true. There is also no evidence that he wants

to be apart from Amanda (5). Finally, answer (3), because he is Christian, is not correct, because according to Amanda, Tom's movie-going is not Christian.

18. **(5)** Tom's job (1), lacks excitement. We know that's why he goes to the movies (see explanation for question 17, above), and Tom specifically says "Adventure is something I don't have much of at work." Therefore (1) is incorrect. We don't know if Tom's job pays well (2) or is a night job (3), so these answers are not correct. Finally, we know that Tom's job does not keep him happy, so answer (4) cannot be correct. There is no evidence that Tom is unfaithful (5).

19. **(2)** Amanda is upset because (2), Tom is always out. Amanda wants to know where Tom goes at night and why he's always at the movies. We learn from their conversation that he goes out because he's not happy, and Amanda becomes upset that he's not happy.

20. **(2)** We can tell from this passage that Tom and Amanda (2), fight often. We know they are not very open with each other (1) because Amanda says, "There's so many things in my heart that I cannot describe to you!" and Tom says the same ("There's so much in my heart that I can't describe to *you*!"). We can infer that Amanda is not adventurous since she tells Tom "Not everybody has a craze for adventure" and tries to tell him to control his instincts, so (3) cannot be correct. This passage shows us that Tom, at least, isn't happy, so (4) cannot be correct. Finally, we know they do not agree on everything so (5) cannot be correct.

21. **(4)** Amanda thinks Tom (4), should be satisfied with what he has. She tells him that people who do not find adventure in their careers "do without it!" She tells him that life "calls for—Spartan endurance!" Furthermore, she is clearly upset by his restlessness and de-

sire for adventure. We know that she wants him to stay home at night instead of going to the movies and that she wants him to control his instinct for adventure, because instinct "belongs to animals!"

Passage D

22. **(3)** Buck's full name is (3), Buck Grangeford. We know that there is a feud between the Grangefords and Shepherdsons, and that Buck shot at Harney Shepherdson. Therefore, we can safely assume that Buck is a Grangeford.

23. **(3)** Buck shoots at Harney because (3), Harney is a Shepherdson. We know that answer (2), Harney insulted Buck, cannot be correct, because Buck tells the narrator that Harney "never done nothing" to him. Harney did have a gun (4), and Harney did shoot at Buck (1), but that was only after Buck shot at Harney first. Furthermore, we know (3) is correct because Buck says he wanted to kill Harney "on account of the feud—because he's a Shepherdson." He knows who Harney is so (5) cannot be correct.

24. **(1)** Miss Sophia turned pale because (1), she was afraid that Harney got hurt. We can tell this is the correct answer because the sentence that tells us "Miss Sophia turned pale" also tells us that "the color come back when she found the man warn't hurt." She is not concerned that she herself would get shot (5). Answer (4), she was angry that Buck shot from behind the bush, is incorrect; the only person who expresses a thought about shooting from behind a bush is Buck's father. There is no evidence that Miss Sophia hates Harney (2), so that answer is incorrect. Furthermore, though she may have been afraid Buck got hurt (3), the sentence quoted above shows us her real concern was for Harney.

25. **(2)** Buck wanted to kill Harney (2), because of a feud between their families. See the explanation above for question 23.

26. **(4)** The narrator's language shows that (4), he has had little formal schooling. The narrator speaks in informal or conversational English rather than formal English. (For example, he says "warn't" instead of wasn't," "We done it" instead of "We did it.") We can assume that a well-educated boy (1) would not speak this way. Furthermore, this is the language of a native speaker, not a foreigner, so (2) is not correct. The narrator may indeed be an orphan (3), but there's no evidence of this in the passage. There is also no evidence that he is a politician (5).

27. **(5)** No one knows (5) what the families started fighting over. Buck tells the narrator that even the old people "don't know now what the row was about in the first place." The best he can do is say "There was trouble 'bout something." The boys speculate that the fight may have been about land (2), but it is clear that they do not know for sure, so (5) can be the only answer.

28. **(1)** This passage shows (1), the stupidity of the feud. It is ironic that Buck is so eager to kill Harney, yet Buck has no idea what he was fighting for. He is simply eager to continue the feud. Furthermore, we can see that the other members of the family—Miss Charlotte, the two young men, and Miss Sophia—don't seem to approve of the feud. Finally, Buck, when explaining the feud, tells the narrator that a feud continues and spreads within the families until "by and by everybody's killed off, and there ain't no more feud." The clear absurdity of this statement—and the seriousness with which Buck says it—supports the notion that the feud is ridiculous.

29. **(2)** In the description of Miss Char-

lotte, "held her head up like a queen" is (2), a simile. The writer compares Miss Charlotte to a queen using the word "like." A metaphor (1) would make the comparison without using "like" or "as." There is nothing ironic about the statement so (5) is incorrect.

Passage E

30. **(3)** Richard Cory was (3), envied by everyone. This is revealed in line 12, when the speaker says that everything about Richard Cory worked "To make us wish that we were in his place." We know that (1), hated by everyone, cannot be correct, because "he was everything" that the "people on the pavement" wanted to be. Answer (2), loved by everyone, is not the most logical answer because the people "wished that [they] were in his place," but that does not mean that they loved him. Finally, (4), looked down upon by everyone, cannot be correct because the people who looked at him were the people "on the pavement"; if anything, they looked up at him, as their envy suggests. He definitely did not go unnoticed so (5) cannot be correct.

31. **(4)** The "we" in this poem is (4), members of the working class. This is clear because the "we," lines 13–14 tell us, could afford only bread, not meat to eat. We know that "we" cannot be (1), members of the upper class or (2), members of the royalty, because they would not envy Richard Cory so much, and they would not have to go without meat. Furthermore, answer (3), slaves, cannot be correct because we have no evidence that these people were enslaved.

32. **(5)** In line 13, "waited for the light" suggests that they were (5), waiting for riches. We know that they could not have been waiting for bread (3), since line 14 tells us they "cursed the bread." The "light" here is a metaphor for the lightness they would feel if they had money, since money would ease many of their heavy burdens.

33. **(3)** Richard Cory shot himself because (3), he had money but not happiness. This is an inference we can make because there is no evidence that (1), he had lost money. We know that (2), he was cursed by the people, is incorrect because they cursed only their bread (their poverty), Richard Cory was a gentleman, and they admired him. Furthermore, we can assume that if he had meat, he also had bread (the poor could only afford the bread). Finally, we can infer that he was not happy because there is no mention of happiness in Richard Cory's life, only riches and manners. We see only a glittering surface. It is ironic because he had a far easier life, it seems, than the working people in town, yet he was obviously so unhappy that he chose to kill himself rather than to continue living. "Went home and put a bullet through his head" suggests it was a very deliberate action, so (5) is incorrect.

34. **(2)** This poem is (2), a ballad. The repeated *abab* rhyme scheme in each stanza and the fact that the poem tells a story, as ballads often do, identifies this as a ballad. A sonnet (1) has only 14 lines, so (1) cannot be correct. Free verse (3) does not have a rhyme scheme, and a limerick (4) is limited to just five lines. A haiku is only three lines, so (5) cannot be correct.

35. **(3)** The theme of this poem is (3), money can't buy you happiness. See the explanation for question 33, above. Note that (5) is incorrect because you can't infer that money itself is the cause of unhappiness.

Passage F

36. **(2)** "Lily Daw and the Three Ladies" is the first story in *A Curtain of Green* because (2), it clearly establishes the theme of the other

stories. The writer tells us that this story "represents most clearly the choice open to white southern women" and "it is for this reason that Welty chose it to open her first collection." Answer (1), it is darker than the other stories, cannot be correct, because the author tells us that it introduces the "other, darker stories" in the collection. We do not know how it compares to the others in length, so (4), it is the longest story, cannot be correct. Furthermore, the writer does not tell us whether or not this is (3), the best story, so that answer cannot be correct. When the story was written (5) has nothing to do with the story's placement.

37. **(1)** Lily is nearly confined in an asylum because (1), she begins to rebel. The writer tells us that Lily "has suddenly begun acting independently, talking back to her elders, sneaking away to circuses, and showing an interest in the opposite sex. The town's matriarchs first decide that confinement in an asylum can be the only remedy."

38. **(4)** The irony of this story is that (4), Lily is more sane than the ladies. This is revealed when the writer tells us that the ladies' "ideal of marriage comes to seem like a kind of confining madness, whereas Lily's unconscious defiance of the community's standards becomes a liberating sanity."

39. **(5)** The message of the story appears to be (2), women should not be forced to conform. This is expressed in the first paragraph, which discusses the restrictions placed on white southern women: "a woman may either be a married and respectable lady…or she may be eccentric like the retarded girl Lily Daw, continually threatened by scandal, madness, and confinement." That the ladies try to force Lily to conform further emphasizes the idea that the author is upset about the limited choices. Furthermore, that Lily's rebellion is seen as madness by the ladies—when Lily is really more sane than they—indicates that rebellion should not be seen as mad.

40. **(1)** The author of this passage thinks that Welty (1), is a gifted writer. His passage is entirely complementary of Welty's skills.

Passage G

41. **(3)** The author shouldn't shoot the elephant because (3), the elephant is a valuable worker. True, he is not the owner of the elephant (1), but the author expressly says that it "is a serious matter to shoot a working elephant—it is comparable to destroying a huge and costly piece of machinery—and obviously one ought not to do it if it can possibly be avoided." There is no indication that it is (2), illegal to shoot elephants. Finally, there is no suggestion that (4), the elephant was no longer able to work now, so this answer cannot be correct either.

42. **(1)** The author was chasing the elephant because (1), it had been acting savagely. The author writes that the elephant's "attack" had passed but that he would "watch him for a little while to make sure that he did not turn savage again."

43. **(4)** A "conjurer" is (4), a magician. This can be determined from the context of the sentence: "They were watching me as they would watch a conjurer about to perform a trick." We can assume that a conjurer is someone who performs tricks, like a magician. Furthermore, he calls his rifle "magical" in the next sentence.

44. **(3)** The author shot the elephant because (3), the people wanted him to. He tells us that "suddenly I realized that I should have to shoot the elephant after all. The people expected it of me and I had got to do it." We

know that (1), it was savage and dangerous, cannot be correct because we know the elephant's attack had passed. Answer (2), because he wanted to, also cannot be correct because the author explicitly tells us that he did not want to: "I did not in the least want to shoot him." Finally, (4), he was ordered to, cannot be correct because there is no indication in the passage that he was ordered by the people to do so. He merely felt the pressure of their expectation; thus, (3) is the best answer. There is no reason to believe he would be killed, so (5) is incorrect.

45. **(2)** This event took place in (2), Asia. We can assume this because the author talks about being a "white man" in front of a "native crowd," and he talks of the "white man's dominion in the East." Furthermore, he tells us that the natives have "yellow faces," and yellow is the color often used to describe the skin of people from Asia. Thus, (2) is the best answer.

Interpreting Literature
and the Arts

Appendix A: Glossary of Terms

INTERPRETING LITERATURE AND THE ARTS

APPENDIX A: GLOSSARY OF TERMS

allegory—a story in which the characters symbolize ideas or values.

alliteration—the repetition of sounds, especially at the beginning of words.

antagonist—the person, force, or idea working against the protagonist.

antihero—a character who is pathetic rather than tragic, who does not take responsibility for his or her destructive actions.

argument poem—a poem that attempts to convince the reader or the person being addressed by the speaker.

aside—in drama, when a character speaks directly to the audience or another character concerning the action on stage; only the audience or character addressed in the aside is meant to hear.

assertion—a statement or declaration, especially one that requires evidence or explanation to be accepted as true.

autobiography—The true account of a person's life written by that person.

ballad—a poem that tells a story, usually rhyming *abcb*.

biography—the true account of a person's life written by someone other than that person.

blank verse—poetry in which the structure is controlled only by a metrical scheme.

characters—people created by an author to carry the action, language, and ideas of a story.

climax—the turning point or high point of action and tension in the plot.

comedy—humorous literature that has a happy ending.

commentary—literature written to explain or illuminate other works of literature or art.

conflict—a struggle or clash between two people, forces, or ideas.

connotation—implied or suggested meaning.

context—the words and sentences surrounding a word or phrase that help determine the meaning of that word or phrase.

couplet—a pair of rhyming lines in poetry.

denotation—exact or dictionary meaning.

denouement—the resolution or conclusion of action.

descriptive essay—an expository essay whose primary purpose is to describe a person, place, or thing.

dialect—language that differs from the standard language in grammar, pronunciation, and idioms (natural speech versus standard English); language used by a specific group within a culture, such as a social class or ethnic group.

dialogue—the verbal exchange between two or more people; conversation.

diction—the particular choice, use, or arrangement of words.

didactic—aiming to teach or instruct (rather than entertain).

drama—literature that is meant to be performed.

dramatic irony—in drama and fiction, when the members of the audience or readers know what more than one or more of the characters know.

elegy—a poem that laments the loss or death of someone or something.

exact rhyme—the repetition of exactly identical stressed sounds at the end of words.

exposition—the setting forth or explaining of ideas or facts; in fiction and drama, the conveyance of background information necessary to understand the plot to be developed.

expository essay—an essay whose primary purpose is to explain or demonstrate.

eye rhyme—words that look like they should rhyme because of spelling, but because of pronunciation do not.

fable—a short story designed to teach a moral lesson or reveal a truth about human nature.

falling action—the events that take place immediately after the climax in which "loose ends" are tied up.

feet—in poetry, a group of stressed and unstressed syllables. In the iambic metrical pattern, a foot consists of one stressed syllable and one unstressed syllable.

fiction—prose literature about people, places, and events invented by the author.

figurative language—comparisons not meant to be taken literally but used for artistic effect (similes, metaphors, and personification).

foreshadowing—a suggestion or indication of things to come.

free verse—poetry that is free from any restrictions of meter and rhyme.

genre—a category or kind; in literature, the different categories of writing.

haiku—a short, imagistic poem composed of three unrhymed lines usually totaling 17 syllables (five, seven, and five syllables per line).

half-rhyme—the repetition of the final consonants at the end of words.

hyperbole—extreme exaggeration not meant to be taken literally, but done for effect.

iambic pentameter—the most common metrical pattern in poetry in which lines have ten syllables and the stress falls on every second syllable.

imagery—the representation of sensory experience through language.

imagistic poem—a poem that seeks mainly to create a vivid image through language.

inference—a conclusion arrived at based upon reason, fact, or evidence.

irony—see dramatic irony or verbal irony.

jargon—technical or specialized vocabulary.

journalism—current information and opinions expressed in newspapers, magazines, and other periodicals, as well as television and radio.

literature—something written or published that is valued for the beauty of its message, form, and emotional impact.

melodrama—a play that starts off tragic but has a happy ending.

mental theater—the "theater" readers create in the mind's eye when reading drama.

metaphor—a type of figurative language that compares two things by saying they are equal (e.g., your eyes are the deep blue sea).

meter—the number and stress of syllables in a line of poetry.

monologue—in drama, a play or part of a play performed by one character speaking directly to the audience.

mystery and morality plays—medieval plays that were dramatic versions of fables and parables.

myth—a story that attempts to explain a cultural custom, practice, or belief or a natural phenomenon.

narrative poem—a poem that tells a story and is driven forward by action.

narrative essay—an expository essay whose primary purpose is to describe an experience or event.

narrator—in fiction, the character or person who tells the story.

non-fiction—prose literature about real people, places, and events.

novella—a very short novel or very long short story.

ode—a poem that celebrates a person, place, thing, or event and expresses profound thoughts about its subject.

onomatopoeia—when the sound of a word echoes its meaning, like "buzz."

parable—a short, allegorical tale that illustrates a moral or religious truth.

paradox—a statement or phrase that seems to contradict itself or to conflict with common sense but which contains some truth.

parody—the imitation of a well-known work for comic effect.

periodicals—publications like newspapers and magazines that are issued at regular intervals.

persona—the speaker or "narrator" of a poem.

personification—figurative language that endows non-human or non-animal objects with human/animal characteristics (e.g., the sun screamed down upon us).

persuasive essay—an expository essay whose primary purpose is to convince or persuade readers.

plot—the ordering of events in a story.

poetry—literature written in verse characterized by its focus on words and images.

point of view—the perspective from which something is told or written.

prose—literature that is not in poetic form (verse) or dramatic form (stage or screenplay).

protagonist—the "hero" or main character of a story, the one who faces the central conflict.

pseudonym—a false name used by a writer.

pun—a play on the meaning of a word.

quatrain—in poetry, a stanza of four lines.

realism—a movement in literature that attempted to represent real life as accurately as possible and that focused on "ordinary" characters dealing with the "ordinary" struggles of everyday life, as opposed to the extraordinary characters and events in tragedies.

reversal—in satire, when two things are switched or interchanged for effect.

rhetorical mode—how an essay is classified based upon its structure, techniques, and purpose (see *descriptive essay*, *narrative essay*, and *persuasive essay*).

rhyme—the repetition of an identical or similar stressed sound(s) at the end of words.

rhymed and metered verse—poetry whose structure is controlled by both a rhyme scheme and metrical scheme.

rhythm—the overall effect of the pattern of words and sentences (length, meter, etc.) in literature, especially poetry.

sarcasm—sharp, biting language intended to ridicule.

satire—a form of writing that exposes and ridicules its subject with the hope of bringing about change.

sestina—a complex free-verse poem of 39 lines that is divided into stanzas of six lines each. The end-words of the six lines in the first stanza must be used as the end-words in the remaining five stanzas but in a different (but predetermined) order. The final three lines must include all six of the end-words from the first stanza.

setting—the time and place in which a story occurs.

simile—a type of figurative language that compares two things using "like" or "as" (e.g., your eyes are as blue as the sea).

soliloquy—in drama, a speech made by a character who reveals his or her thoughts to the audience as if he or she is alone and thinking aloud.

sonnet—a poem composed of fourteen lines, usually in iambic pentameter, and often rhyming; *abba abba cdcdcd* (Italian) or *abab cdcd efef gg* (Shakespearean).

stage directions—in a play, the instructions provided by the playwright that explain how the action should be staged, including directions for props, costumes, lighting, blocking, and tone.

stanza—a group of lines held together in a poem, like a paragraph in an essay.

structure—the manner in which a work of literature is organized, its order of arrangement and divisions.

sub-genre—a category within a larger category (e.g., science fiction is a sub-genre of fiction).

suspense—the state of anxiety caused by an undecided or unresolved situation.

symbol—a person, place, or object invested with special meaning to represent something else (e.g., a flag).

theme—the overall meaning or idea of a work of fiction, poetry, or drama.

thesis—the main idea of a non-fiction text.

thesis statement—the sentence(s) that express the author's thesis.

tone—the mood or attitude conveyed by writing or voice.

topic sentence—the sentence in a paragraph that expresses the main idea of that paragraph.

tragedy—a play that presents a noble character's fall from greatness due to a tragic flaw.

tragic hero—the character in a tragedy who falls from greatness and accepts responsibility for that fall.

tragic flaw—the characteristic of a hero in a tragedy that causes his or her downfall.

tragicomedy—a tragic play that includes comic scenes to balance the sadness.

understatement—a statement that is deliberately restrained.

verbal irony—when the intended meaning of a word or phrase is the opposite of the expressed meaning.

villanelle—a complex rhymed and metered poetic form composed of five stanzas of three lines, each rhyming *aba* and a final quatrain rhyming *abaa*; in addition, line one is repeated in lines six, 12, and 18, and line three is repeated in lines nine, 15, and 19.

visual poem—a poem whose physical form reflects its meaning.

voice—in non-fiction, the sound of the author speaking directly to the reader.

wit—expressing keen observations in an amusing or unusual way.

Interpreting Literature and the Arts

Appendix B: Prefixes, Suffixes, and Word Roots

INTERPRETING LITERATURE AND THE ARTS

APPENDIX B: PREFIXES, SUFFIXES, AND WORD ROOTS

Prefix	Meaning	Example	Definition
uni-	one	unity	state of being one, oneness
mono-	one	monologue	a long speech by one person
bi-	two	biweekly	every two weeks
duo-	two	duality	having two sides or parts
tri-	three	triangle	having three angles
quadri-	four	quadruped	an animal with four feet
tetra-	four	tetrad	a group or set of four
quint-	five	quintuplets	five offspring born at one time
pent-	five	pentacle	a star with five points
multi-	many	multifaceted	having many sides
poly-	many	polyglot	one who speaks or understands several languages
omni-	all	omniscient	knowing all
micro-	small	microcosm	little or miniature world
mini-	small	minority	small group within a larger group
macro-	large	macrocosm	the large scale world or universe
ante-	before	antebellum	before the war
pre-	before	precede	to come before
post-	after	postscript	message added after the close of a letter
inter-	between	intervene	to come between
inter-	together	interact	to act on or influence each other

Prefix	Meaning	Example	Definition
intra-	within	intravenous	within the vein
intro-	into, within	introvert	person withdrawn into him/herself
in-	in, into	induct	to bring in (to a group)
ex-	out, from	expel	to drive out or away
circum-	around	circumscribe	to draw a line around
sub-	under	subvert	to undermine
super-	above, over	supervisor	one who watches over
con-	with, together	consensus	to agree with, general agreement
non-	not	nonviable	not able to live or survive
in-	not	invariable	not changing
un-	not, against	unmindful	not conscious or aware of
contra-	against	contradict	to speak against
anti-	against, opposite	antipode	exact or direct opposite
counter-	against, opposing	counterproductive	working against production
dis-	not, away, opposite of	dispel	to drive away
		disorderly	not having order
mis-	wrong, ill	misuse	to use wrongly
mal-	bad, wrong, ill	maltreat	to treat badly or wrongly
		malaise	feeling of discomfort or illness
pseudo-	false, fake	pseudonym	false or fake name

Suffix	Meaning	Example	Definition
-en	to cause to become	broaden	to make more broad, widen
-ate	to cause to be	duplicate	to make or be an exact copy
-ify/-fy	to make or cause to be	electrify	to charge with electricity
-ize	to make, to give	alphabetize	to put in alphabetical order
-al	capable of, suitable for	hysterical	characteristic of hysteria
-ial	pertaining to	artificial	not made by nature, made by man

Suffix	Meaning	Example	Definition
-ic	pertaining to	aristocratic	pertaining to the aristocracy
-ly	resembling, having the qualities of	fatherly	like a father
-ly	in the manner of	boldly	in a bold manner
-ful	full of	meaningful	full of meaning
-ous, -ose	full of	wondrous	full of wonder
-ive	having the quality of	descriptive	having a good description
-less	lacking, free of	loveless	lacking love
-ish	having the quality of	impish	having qualities of an imp
-ance/ence	quality or state of	expectance	state of waiting, expecting
-acy	quality or state of	expectancy	in a state of expectance
-or/er	one who does or performs action of	singer	one who sings
-arium/orium	place for	aquarium	place for water
-ary	place for, pertaining to	sanctuary	a sacred place, refuge
-cide	kill	pesticide	substance for killing insects
-ism	quality, state or condition of; doctrine of	environmentalism	doctrine of protecting the environment
-ity	quality or state of	duality	having two sides
-itis	inflammation of	tonsillitis	inflammation and infection of tonsils
-ment	act or condition of	judgment	act of judging
-ology	the study of	zoology	study of animals

COMMON LATIN WORD ROOTS

Root	Meaning	Example	Definition
equus	equal	equilateral	triangle with equal sides
amare	to love	amorous	readily showing or feeling love

Root	Meaning	Example	Definition
audire	to hear	auditory	of the hearing
capere	to take	captivate	to capture the fancy of
dicere	to say, speak	dictate	to state or order; to say what needs to be written down
duco	to lead	conduct	to lead or guide (through)
facere	to make, to do	manufacture	to make or produce
lucere	to light	lucid	very clear
manus	hand	manicure	treatment of the fingernails
medius	middle	median	middle point
mittere	to send	transmit	to send across
omnis	all, every	omnipresent	present everywhere
plicare	to fold	application	putting one thing on another
ponere/positum	to place	position	place a person or thing occupies
protare	to carry	transport	to carry across
quarere	to ask, question	inquiry	investigation, questioning
scribere	to write	scribe	person who makes copies of writings
sentire	to feel	sentient	capable of feeling
specere	to look at	spectacle	striking or impressive sight
spirare	to breathe	respiration	act of breathing
tendere	to stretch	extend	to make longer, stretch out
verbum	word	verbatim	word for word

COMMON GREEK ROOTS

Root	Meaning	Example	Definition
bios	life	biology	study of life
chronos	time	chronological	arranged in the order in which things occurred
derma	skin	dermatology	study of skin
gamos	marriage, union	polygamous	married to many at once

Root	Meaning	Example	Definition
genos	race, sex, kind	genocide	deliberate extermination of one race of people
geo	earth	geography	study of the Earth's surface
graphein	to write	calligraphy	art of beautiful handwriting
krates	member of a group	democrat	member of the democratic party
kryptos	hidden, secret	cryptic	concealing meaning, puzzling
metron	to measure	metronome	device with a pendulum that beats at a determined rate to measure time
morphe	form	polymorphous	having many forms
pathos	suffering, feeling	pathetic	arousing feelings of pity or sadness
philos	loving	philosophy	love of, or the search of, knowledge or wisdom
phobos	fear	acrophobia	fear of heights
photos	light	photolysis	chemical decomposition due to the action of light
podos	foot	bi-pedal	having two feet
psuedein	to deceive	pseudonym	a false name
pyr	fire	pyrotechnics	fireworks display
soma	body	psychosomatic	involving mind and body
tele	distant	telescope	instrument to see objects a great distance away
therme	heat	thermos	jug or bottle that keeps liquids hot

Available at your local bookstore or order directly from us by sending in coupon below.

RESEARCH & EDUCATION ASSOCIATION
61 Ethel Road W., Piscataway, New Jersey 08854
Phone: (732) 819-8880

☐ Payment enclosed
☐ Visa   ☐ MasterCard

Charge Card Number

Expiration Date: _____ / _____
                    Mo   Yr

Please ship REA's **"GED"** @ $17.95 plus $4.00 for shipping.

Name _____

Address _____

City _____ State _____ Zip _____